Guide to the Geology of Mesa Verde National Park

Mary O. Griffitts

Printed by Lorraine Press, Utah
International Standard Book Number 0-937062-11-1
Library of Congress, Catalogue Card No. 88-063410
© ℗ 1990 by Mesa Verde
Museum Association, Inc.
Mesa Verde National Park, Colorado 81330

Contents

Illustrations

ERA	MYA*	PERIOD	FORMATION	DESCRIPTION OF ROCKS	ENVIRONMENT OF DEPOSITION
M E S O Z O I C	78	C R E T A C E O U S	Cliff House	Sandstone; clams, snails, ammonites, shark, fish.	Marine, transgressive, near shore.
	81		Menefee	Dark shale, coal, sandstone, bentonite; abundant plant remains	Continental, stream and swamp.
			Point Lookout	Sandstone	Marine, mostly regressive near shore.
	93		Mancos	Dark shale, thin sandstone, limestone, bentonite; ammonites, clams, oysters, fish, shark, rays.	Marine, mostly transgressive, offshore
	100		Dakota	Sandstone, dark shale and coals.	Marine transgressive, near shore
	138	?	Burro Canyon	Conglomerate	Continental, rapid stream
		JURASSIC	Morrison	Shale, sandstone, limestone; dinosaurs, other reptiles, primitive mammals.	Continental, swamps, broad streams
			Summerville	Shales, sandy shales	Continental, inland lake
	205		Entrada	Reddish orange, fine sandstones	Continental, sand dunes
		TRIASSIC and/or JURASSIC	Dolores	Sandstone, conglomerate; reptile teeth	Continental, stream, lake, flood plain
	240		Navajo	Orange-red sandstone	Continental, mostly windblown
P A L E O Z O I C	290	PERMIAN	Cutler	Red-brown sandstone	Continental, flood plain
	330	PENN.	Hermosa	Limestone and salt	Marine, very shallow
	360	MISS.	Leadville	Limestone, corals, crinoids, brachiopods.	Marine, offshore
		DEVONIAN	Ouray	Limestone, massive; many invertebrates	Marine, offshore
			Elbert	Shale, thin limestones, sandstones, green to red	Marine, tidal flat
	380		Aneth	Light gray to red sandstone	Continental, flood plain
	500	NO ORDOVICIAN OR SILURIAN ROCKS KNOWN - EROSION INTERVAL			
		CAMBRIAN	Ignacio	Quartzite, quartzose sandstone, shale lenses	Marine, shallow water
	570+	PRECAMBRIAN		Quartzites, schists	Marine

— — Unconformity, erosion interval

Figure 1: Geologic column at Mesa Verde National Park.

MYA* = Millions of years ago

ERA	PERIOD	MYA*	IMPORTANT LIFE FORMS
CENOZOIC	QUATERNARY	.01	Modern man.
		2	Mammoths, mastodons, modern mammals.
	TERTIARY		Rise of modern mammals, invertebrates, birds; dinosaurs and ammonites extinct.
	LARAMIDE REVOLUTION	63	
MESOZOIC	CRETACEOUS		Mammals expanding, advent of flowering plants, ammonites abundant, dinosaurs abundant.
		138	
	JURASSIC		Ammonites maximum, giant dinosaurs, first birds.
		205	
	TRIASSIC		First dinosaurs, primitive mammals, conifers.
		240	
PALEOZOIC	PERMIAN		Reptiles.
		290	
	PENNSYLVANIAN		Primitive reptiles, large amphibians, insects, ancient plants.
		330	
	MISSISSIPPIAN		Criniods, amphibans.
		360	
	DEVONIAN		Shark like fishes, armored fishes, primitive land plants.
		410	
	SILURIAN		First backboned animals, first known land plants.
		435	
	ORDOVICIAN		Culmination of trilobites, rise of corals and other invertebrates.
		500	
	CAMBRIAN		Rise of shell bearing molluscs, dominance of trilobites, well developed marine faunas of invertebrates.
		570	
	PRECAMBRIAN		Primitive marine animals; algae.

MYA* = Millions of years ago.

Figure 2: Geologic time table.

6

Introduction

Mesa Verde National Park was established in 1906 to protect and study the large concentration of Anasazi Indian sites on the mesa tops, cliffs and canyons. Although most of the thousands of visitors to the Park are attracted by the archeological sites, the spectacular scenery certainly enhances the enjoyment of the area. This book provides first, a road log with brief descriptions of the geology at many view points. The second section contains a more detailed geologic history of the region from some two billion years ago to the present, along with an explanation of some of the basic geological processes at work. References are given throughout the road log to more detailed discussions of specific topics in the second section of the text. A glossary at the end of the book may help with unfamiliar terms.

The road log is divided into five sections: unit 1 — from the entrance station to Park Headquarters on Chapin Mesa, unit 2 — Spruce Tree Canyon at Park Headquarters, unit 3 — the loop road to Cliff Palace and Balcony House, unit 4 — the loop road to the mesa top ruins, and unit 5 — Wetherill Mesa Road. Road log stops are shown on the geologic map (Plate A) which supplements the National Park map.

A chart (Figure 1) summarizes the total geologic column in the Mesa Verde area. Although only the upper four formations, Mancos through Cliff House (shown in **bold type**), are exposed on the surface within the Park, the relative positions of the older, underlying rocks contribute to a more complete understanding of the regional geology. The geologic time table (Figure 2) should help put the history in time perspective.

Acknowledgments

In preparing this guidebook the author has drawn freely from published geologic literature. A limited bibliography is presented at the end of the book. Personal consultation with a number of experts in specialized fields has been of great assistance. The author is especially grateful to William S. Cobban of the United States Geological Survey who has provided much information on the invertebrate fossils. Jack Wolfe, also of the United States Geologic Survey, identified the plant fossils. Kenneth Carpenter of the University of Colorado Museum of Paleontology identified the vertebrate fossils. To these experts the author wishes to express her appreciation for the time and expertise involved. The interpretation of this information and use within this present context must be the sole responsibility of the author.

The National Park Service has made the field and museum work possible through the V.I.P. program. Although all the Mesa Verde National Park personnel have been most helpful and encouraging, special thanks are expressed to Superintendent Robert Heyder who approved the study, Chief of the Research Laboratory Dr. Jack Smith and Museum Curator Allen Bohnert for providing space to work, equipment, access to collections, location information, and above all, interest and encouragement. Marilyn Colyer's detailed knowledge of the geography and natural history of the Park and interest in the project have been especially helpful. Chief of Stabilization Kathy Fiero and other members of the Stabilization Crew have helped in locating and bringing in specimens. Chief of Interpretation Don Fiero has given generously of his time in assisting with photography, reviewing the manuscript and encouraging the writer. Without help from these and many other Park personnel, this work would not have been accomplished.

Part I

Road Log
The road log stops are marked on the Geologic Map (Plate I).

Unit 1. Entrance Road to Park Headquarters.
Stop A. Entrance to Mesa Verde National Park, parking lot for trailers.

The parking lot is located near the middle of the Upper Cretaceous Mancos Formation (page 42). The town of Mancos can be seen about seven miles to the east with the rugged La Plata Mountains on the horizon. The Mancos Formation was originally described from the Mancos River Valley in 1899 and consists of about 2000 feet of gray shale, limestone, and thin sandstone deposited in the great sea which completely divided North America (Figure 25, page 41) about ninety million years ago. To the north just outside the Park boundaries are low hills capped by the rusty brown Juana Lopez Member of the Mancos Formation (page 47). The small hills just west of the parking lot are underlain by soft gray shales with thin sandy limestone layers. These shales are very thin-bedded and soft, break easily and make very steep, slippery cliffs. These Niobrara equivalent rocks of the Mancos Formation (page 50) were deposited in the quiet sea, far from shore. The route into the Park lies to the south (Figure 3). The flat land is broken by small hills, most of which are capped by yellow rock. Except for the yellow cap rock, this is entirely middle Mancos Formation, soft, light to dark gray shales and a few very thin sandstones. The yellow rock at the tops of the small hills is slump rock from the overlying Point Lookout Formation. It is not in place. Farther to the south is Point Lookout. This erosion remnant stands some 1400 feet above the parking lot altitude of 7000 feet, and is capped by the Point Lookout Formation (page 51) about five million years younger than the rocks here.

All the rocks in the Park dip very gently three to seven degrees to the south, so, progressing into the Park, one keeps crossing to younger and younger rocks. A cross section (Figure 4) gives an idea of the relationship of the present surface to the whole sequence of rocks in the Park.

In the Park, the road gradually climbs into the upper part of the Mancos Formation. The rock in the road cuts is mostly gray shale, thin bedded, sometimes called paper shale because the layers are so thin. The road is cut along the north and northeast face of Point Lookout in these

Figure 3. View south towards Point Lookout. Broken line indicates approximate contact between Mancos Formation below and Point Lookout Formation above.

very soft shales. This material makes a very difficult road to maintain. The road bed has to be well stabilized below to try to prevent caving, while constant slipping and sliding of the soft shales above brings down loose rock or mud slides onto the surface of the road with every storm. Within the gray shales are occasional thin orange layers. These are weathered bentonites, an altered volcanic ash which may swell when it gets wet, and also causes instability to the road cuts.

The road climbs fairly steeply as it skirts around the Point, climbing into the Upper Mancos shales (page 50). The rocks in the road cuts are lighter colored, slightly yellowish with more sandy layers.

Stop B. Mancos Valley Overlook

A display at the edge of the canyon just east of the parking area describes the major features here. The overlook is close to the contact between the Mancos Formation below and the Point Lookout Formation above. The rocks are light colored, yellowish, sandy shale and sandstone. The great sea in which the Mancos was deposited had started to retreat to the northeast, and more sandy materials were deposited in a near shore environment. The top of Point Lookout above this overlook to the west is capped with the solid sandstones of the Point Lookout Formation (page 51). The mountains on the horizon are the La Platas. These mountains were formed during the Laramide Revolution (page 64) at the end of the Mesozoic era nearly 70 million years ago as the result of laccolithic intrusions. Hot liquid rocks bowed up and squeezed between layers of the pre-existing sedimentary rock and formed large mushroom shaped igneous bodies. Erosion has removed the overlying softer sediments and cut the more resistant igneous rocks into a sharp rugged mountain range. To the east is the high flat erosion surface which gradually rises to the mountain range. This is the remains of the ancient erosion surface of very early Tertiary time (page 65).

Just across Mancos Canyon to the east is a very level bench along the canyon wall. This bench is an old stream terrace developed when the stream stopped cutting downward and began to cut a broader stream bed. The period of more sluggish stream action was interrupted by another uplift of the land, and the stream again began active downward cutting through the old stream bed. This terrace is the remains of the edges of the old stream bed. Such terraces are visible in many of the

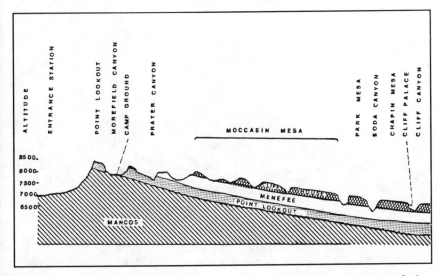

Figure 4. Cross section from Park entrance to Cliff Canyon showing relation between surface and underlying rock.

canyons in the Park. Also across on the east canyon wall is a volcanic plug. It is more resistant to erosion than the Mancos shale into which it was intruded and so stands out as a prominent hill. Small dikes are found extending from the plug. These volcanics are probably between 27 and 30 million years old.

Stop C. Morefield Canyon.

Morefield Canyon has been cut through the hard Point Lookout sandstone cap rock, and the stream has cut a broad valley in the soft Mancos shale. The tops of the ridges surrounding the valley are all capped with resistant Point Lookout Sandstone. Lone Cone is a small erosion remnant called a butte. It was part of the mesa until it was isolated by erosion.

The road follows Morefield Canyon to a tunnel through the divide between Morefield and Prater Canyons.

Figure 5. Knife Edge from Montezuma Valley Overlook (Stop D). Point Lookout sandstone (PL); Mancos Formation (M).

Stop D. Montezuma Valley Overlook.

The very sheer cliff to the north at this stop is known as the Knife Edge. Capped by resistant Point Lookout Sandstone, the lower section is soft gray shale of the upper Mancos Formation (Figure 5). The old road went around Knife Edge for many years, but this section of the road was so hard to maintain because of the caving soft shale that, in 1957, the tunnel was built through the divide between Prater and Morefield Canyons and the old road abandoned. Natural weathering and erosive forces are rapidly removing the traces of the old road which formerly looped around Knife Edge and back into Morefield Canyon above the southwest side of the present campground.

From this point there is an excellent view to the west. The city of Cortez lies on a flat erosion plain and is underlain by Dakota Sandstone (page 41), the first deposit in the great spreading Cretaceous sea. Lying on this flat surface are the lower Mancos shales, and many of the very small hills are capped by the lower Mancos (Greenhorn) limestone (page 44). Erosion has removed more than 3000 feet of younger sediment from the plain and erosion is still working to wear down the Mesa Verde to this lower level. As soon as the cap rock is removed, the underlying shales are rapidly washed away.

The red soil on this flat plain is the result of red dust blown in from the southwest probably over the last million years. The wind blown dust is called loess and forms a rich soil which holds moisture well.

Figure 6. Ute Mountain on the horizon from the North Rim.

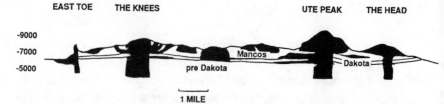

Figure 7. Diagram of geology of Ute Mountain laccoliths, dark shading represents igneous intrusives. (Based on data from Ekren and Houser.)

The view farther to the west (Figure 6) shows the Sleeping Ute Mountain on the horizon. This small mountain range was formed as the result of molten rock being forced upward through the overlying Paleozoic and Mesozoic sediments. Some of the molten material spread out and bowed up the upper Mancos Formation to form mushroom shaped intrusions called laccoliths. Ute Peak is one of these laccoliths (Figure 7). Erosion has removed the bowed up late Cretaceous sediments from the upper part of the mountain, but it is possible to see parts of the Mancos Formation bent upward around the edge of the igneous bodies. Several of these laccoliths, together with dikes and other igneous bodies forming the Sleeping Ute Mountains, were intruded into these pre-existing sediments during the time the modern Rocky Mountains were being formed during the Laramide Revolution some 70 million years ago (page 64).

The road now continues along the west side of Prater Canyon. Across the canyon to the north is an alcove (Figure 32, page 52) which once sheltered a very small cliff dwelling in the Point Lookout sandstone.

Stop E. Park Point.

Between the Montezuma Overlook and Park Point, the road climbs slowly upward through the Point Lookout Formation and the Menefee Formation to the Cliff House Formation. The road cuts change from the pale buff yellow sandstone of the Point Lookout through the dark gray shales and coals of the Menefee to the yellow brown sandstones of the Cliff House Formation — the youngest sandstone of the Park.

Much of the road between Montezuma Overlook and the Park Point turnoff is in the Menefee Formation. Parts of this section of the road have had to be rerouted because the soft woody shales and thin-bedded brittle coal layers of the Menefee Formation were so unstable it was difficult to maintain the road safely.

The sea that deposited the underlying formations (Mancos and Point Lookout) had withdrawn from this particular area, and the great flat sea bed was now slightly above sea level. Muds and sands were deposited along broad sluggish streams. Muddy swamps received large amounts of decaying plant material and gave rise to coal beds and brown woody shales. The coal beds here are too thin to be mined, but outside the Park are thicker, mineable coals. Some of the layers of coal burned at some time in the past, possibly because of lightning strikes, and baked the overlying shale to a red color. Pieces of this burned red shale were shaped and used as ornaments by the Anasazi.

A foot trail leads to the observation point where there is a pano-ramic view of the surrounding country. To the east on the horizon are the La Plata Mountains, a laccolithic mountain range (see Stop B for description), formed during the Laramide Revolution (page 64). The

Figure 8. Park Point view northeast, La Plata Mountains on horizon, small fault marked with broken line dropped the north end of the spurs about 50 feet. Knife Edge (KE); typical rapidly eroding soft Mancos shale (M); Point Lookout Sandstone (PL).

Figure 9. Park Point, view west of Sleeping Ute Mountain (See Figure 7 for diagram.) Spurs along North Rim broken by joints and faults marked by arrows. Mancos Formation (M); Point Lookout Sandstone (PL); Menefee Formation (MF); Cliff House Formation (CH).

broad flat plain below the Mesa Verde is floored with Mancos and Dakota deposits. The resistant Point Lookout Sandstone caps the mesa at Knife Edge, and the typical erosion pattern in the soft Mancos shales shows at the base of the cliffs. In the foreground a small fault cuts off the north ends of the spurs (Figure 8), dropping the Menefee coal beds about 50 feet down against the Point Lookout Sandstone. Many of the ends of the spurs along this northern edge of the mesa are cut by nearly east-west trending small faults and joints which control the erosion pattern along the North Rim (Figure 9).

The San Juan Mountains in the distance to the north are the eroded remnants of volcanic rocks about 35 million years old. Many of the low buttes on the plain are capped by the Greenhorn limestone member of the Mancos Formation(Page 44).

To the west (Figure 9) the spurs of the Mesa Verde North Rim are deeply eroded by headward erosion of small streams. Almost all the spurs show the entire sequence of sediments seen in small segments elsewhere in the Park. The lowest part of the slopes are the gray shales of the Mancos Formation deposited in the great advancing sea. The sandstones overlying the soft dark shales are the Point Lookout Formation, the near shore sediments deposited as the sea retreated a little way to the northeast. These sandstones form strong cliffs and bare rock surfaces. The sea withdrew from the area, and overlying these strong sandstones are the softer deposits of the Menefee Formation (page 53), shales, coal beds and soft sandstones which were deposited on the great, flat, former sea bottom in swamps, lagoons and broad flat sluggish stream beds. The sea again advanced over the area, and the brown sandstones of the Cliff House Formation (page 58) were deposited.

The silhouette of Sleeping Ute Mountain on the horizon to the west suggests the outline of a man lying on his back with his head to the north and arms crossed over his chest. The origin of this small mountain range is discussed under Road Stop D (page 13).

To the south is a good overall view of the flat top of the mesa, sloping gently to the south and cut by sharp shouldered canyons. On a clear day, Shiprock can be seen over 30 miles to the south. Shiprock is a volcanic neck with radiating dikes formed in mid Tertiary time, about 30 million years ago.

Stop F. North Rim Overlook.

A complete sequence of the sedimentary rocks of Mesa Verde National Park is visible from this location (Figure 10). To summarize what is described in more detail in other parts of this book, the floor of the broad valley below the mesa is in the Dakota Sandstone, about 100

Figure 10. View north from North Rim Overlook showing Mancos Formation (M); Point Lookout Sandstone (PL); Menefee Formation (MF); Menefee sandstone (MFS); Cliff House Formation (CH); Landslide scar (L); Fault Scarp (FS).

million years old, the first deposit of the great inland sea which covered the continent from about the present Mississippi Valley to western Utah, and from the Arctic Ocean to the Gulf of Mexico (Figure 25, page 41). As the sea advanced and deepened, the shallow water sand deposits gave way to deeper water mud deposits. These became the Mancos shales seen in the lower parts of the cliffs as soft gray and tan shales. Small hills on the valley floor are also remnants of the Mancos shales. The sea reached its maximum extent and began a very slow, pulsating retreat. Two thousand feet of the Mancos Formation were deposited here in the approximately 10 million years that the quiet sea covered this area. The sea began to retreat, and the Point Lookout Sandstone which forms the strong, light colored cliffs capping the mesas was deposited as barrier beaches and near shore sands. The sea completely withdrew from this location temporarily, and the broad, almost flat, former sea bottom became land. In the warm, humid climate, land plants began to grow, and eventually swamps and lagoons filled with decaying vegetation, which gave rise to the woody brown shales and coal of the Menefee Formation visible in the road cut just north of this stop. Streams drained the area, and the stream beds are marked by the irregular sandstone beds visible in the canyon wall across from here as small, white, discontinuous cliffs in the slope above the strong marine Point Lookout Sandstones. Most of the Menefee Formation is soft dark shales and coal, and forms gentle slopes. The sea readvanced over the area, and the buff to orange sandstones of the Cliff House Formation were deposited. These can be seen capping the mesa to the southwest and in the road cut. Here, along the North Rim, the Cliff House Formation is only about 70 feet thick. A few sandstone beds up to ten feet thick occur in the lower part of the formation here, but it is mostly thin-bedded soft sandstone and shaly sandstone. Compare this section of soft sandstones with the massive Cliff House Formation where the cliff dwellings are found in the canyons along the Ruins Road about eight miles to the south. There, this formation totals 200-300 feet, and individual massive sandstones 50-100 feet thick. This variation suggests that the North Rim was far enough from the shore that it was not receiving large amounts of sand from the land to the southwest. The top layers of these Cliff House Sandstones in both locations bear fossils which are dated at about 78 million years old.

Across the canyon, a sharp diagonal cliff face marks the trace of a fault scarp in the Point Lookout Sandstone. It marks the surface of a break (fault) in the rocks along which the north portion dropped down about 20 feet in relation to the south part. There are a number of these small faults present along the North Rim of the Park. Others are clearly visible at Park Point.

Several landslides are visible on the north side of the canyon. Some of these probably began after extra heavy snows in the fall of 1972 and have continued intermittently with new large slides appearing in 1980. The lower slopes of soft Mancos shales and the upper slopes of soft Menefee shales are very unstable. Once the slides have denuded the steep slopes of trees, continued erosion is rapid, and the slides progressively enlarge.

Stop G. Far View Visitor Center.

Far View Visitor Center, at a little over 8000 feet in elevation, offers another panoramic view of the entire Mesa Verde. The gentle dip slope surface of the high mesa and the abrupt scarp down to the low lying surrounding plain is visible from this overlook. From here, the main road follows the north-south trending Chapin Mesa about four miles to the Park Headquarters where the elevation is about 7000 feet. This road and the main ruins road loops are all on the Cliff House Formation (page 58), the youngest Cretaceous formation in the Park. Most of the Anasazi ruins are located in or on this formation.

Unit 2. Park Headquarters.
Spruce Tree Canyon.

The features visible from the west edge of Spruce Tree Canyon will serve as an introduction to the geology of Chapin Mesa. Not all of the features visible here will be present at each of the other stops on the Ruins Loop Roads. New features will be described as they appear.

The large Spruce Tree House cliff dwelling occupies a deep alcove in the Cliff House Formation. A more detailed description of the formation begins on page 58. Generally, however, in the south part of the Park, the Cliff House Formation consists of two massive, thick, yellow-orange sandstone units separated by a softer sandy shale section. Only the upper sandstone unit is visible at this stop because this part of the canyon is less than 100 feet in depth. The sandstones were laid down as near shore sand bars and sheet sands during the last great advance of the Upper Cretaceous sea nearly 80 million years ago. The sea advanced quite rapidly (geologically speaking) from the northeast, and great quantities of sand were washed down from the lands rising far to the southwest. The cliff forming sandstone is not an uniformly homogeneous deposit. There are thin shaly breaks visible as narrow dark lines in the rock walls. These breaks in the sandstones are responsible for the formation of the larger alcoves occupied by the cliff dwellings.

A more detailed description of the mechanism of alcove formation begins on page 69. Briefly, surface water, rain and melting snow seep into the porous sandstone. The water moves fairly freely until it gets to an impervious layer of rock such as shale. Because it cannot percolate through the shale, the water moves along the top of the shale bedding plane, down-slope. Eventually it comes out at the edge of a canyon. The seepage water dissolves the cement holding the sand grains together, and the sandstone crumbles at the edge of the canyon just above the shale layer. Wind and rain wash away the loose sand grains, and a little groove or niche is formed. Over hundreds of thousands of years this undercut enlarges through a continuation of the seepage solution plus wind and rain erosion to form an alcove.

Along the trail to Spruce Tree House the surface of the rock on the right is weathered into a polygonal pattern of grooves sometimes called "turtle backs" (Figure 51, page 76). The origin of this erosion pattern is not well understood, but it seems to be the result of a combination of conditions — the character of the rock, the exposed position and chemical and mechanical weathering agents.

Down the trail on the left is a smooth sandstone surface on which are numerous dark masses called concretions (Figures 48 and 49, page 73 and 74). Some look like pipes or long rods, some are hollow, others solid. These concretions (page 72) are deposited by ground water moving through the sandstone until it comes to a spot which has a different porosity. The water has mineral matter such as lime and iron dissolved in it, and when anything happens to obstruct the free movement of the water, the minerals are precipitated. The shape of the concretion depends on the shape of the obstruction or area of different porosity, and therefore an almost unlimited variety of shapes is possible. Since these concretions are the result of secondary deposition of extra cement, they are more resistant to weathering than the surrounding sandstone and so remain positive features as the sandstone wears away.

Figure 11. Sandstone-shale contact with small niche developing along trail to Spruce Tree house: sandstone (SS); shale (SH).

On the right near the gate is an early stage of the formation of an alcove (Figure 11). Water seeps out along the thin shale at the base of the overhang and continuously disintegrates the sandstone. Across and toward the head of the canyon, high up, are other small alcoves with a few walls visible in a higher level of the sandstone.

The head of the canyon is very abrupt, a typical box canyon. This is a common type of canyon in arid regions, caused by rapid runoff and vigorous erosion during storms, accompanied by slow but continuous undermining by seeps and spalling such as described with alcove formation.

Spruce Tree Canyon is a very young stream valley with a sharp V-shaped canyon. Most of the erosion is due to undercutting and collapse, or to the very vigorous downcutting during heavy rains and spring runoff.

Figure 12. Cliff Palace in Cliff Canyon. Upper Cliff House sandstone (UCH); middle shaly sandstone break (S); Lower Cliff House sandstone (LCH).

Unit 3. Ruins Loop Road To Cliff Palace and Balcony House.
Cliff Palace

This canyon is cut much deeper than at Spruce Tree House. The cliff dwelling is in the upper part of the Cliff House Formation (Figure 12) just as at Spruce Tree House. Here the canyon is about 240 feet deep and has cut on down through not only the upper massive cliff forming sandstone, but also through the middle sloping shaly sandstone and through the lower Cliff House vertical faced sandstone. The major alcove sheltering Cliff Palace is at about the same level in the upper sandstone as at Spruce Tree House. Directly across Cliff Canyon, remains of other buildings are seen in small alcoves a little younger in the geologic section.

The vertical dark streaks on the upper lips of the alcoves are known as desert varnish. Although the exact cause of desert varnish is not agreed upon by all geologists (page 74), it is now considered to be the result of a combination of processes. In the presence of moisture, bacterial action fixes manganese and iron particles brought in with wind blown dust to the rock faces. It is not known how long this process requires, but it is generally conceded that the darker the deposit, the longer it has been forming.

Down the trail, just before the ruin, the contact of the porous sandstone and impermeable shale can be seen at the left of the trail. The undermining of the sandstone is forming another niche. Evidence of the marine origin of the rocks can be seen within Cliff Palace in the ripple marks and fossils in some of the steps and building blocks.

All along this loop road, soft sandstones and sandy shales of the upper Cliff House Formation lie above the massive cliff forming member.

Fossils in these upper sandstones have been dated at about 78 million years old.

House of Many Windows.

On the floor of this overlook an interesting pattern of solution depressions has developed similar to those shown in Figure 46, page 71. Rain water falling through the atmosphere forms a very weak acid which dissolves the lime cement of the sandstone. Rain water runs off the relatively flat surface at the top of the cliff but accumulates in any slight depressions on the surface. Solution of the limy cement by the water loosens the sand grains so they can blow away when the area is dry or

Figure 13. View north up Soda Canyon from Balcony House where the upper massive Cliff House sandstone gradually becomes more shaly and less resistant to erosion.

wash away with a heavy rain, and the central depression deepens. Little grooves leading to the central depression begin to form as the run off water dissolves the areas of softer cement. Eventually an almost flower-like pattern is formed on some of the cliff tops. Such a pattern is found at Sun Temple where it has been partially enclosed with small walls by the Anasazi (Figure 47, page 72). Another such solution depression is visible near the upper gate on the entrance trail to Balcony House.

Balcony House.

To the north up Soda Canyon (Figure 13) the character of the upper Cliff House sandstone changes. From this region north-eastward, the upper rocks of the Cliff House Formation are no longer a massive cliff forming unit,

Figure 14. Seep at contact of Cliff House sandstone and shale on Balcony House trail.

Figure 15. Balcony House in upper Cliff House sandstone from Soda Canyon overlook.

but have weakened to a less resistant sequence of alternating thinner bedded sandstones and sandy shales. This change in characteristics of the rocks from a massive sandstone to shaly sandstone and sandy shale marks a transition from shallow, near-shore to slightly deeper water deposits as the sea moved slowly southwest. The trail leads down to the shale-sandstone contact where seeps are visible in small alcoves (Figure 14). Balcony House (Figure 15) is in a large alcove in the upper Cliff House sandstone, as are Cliff Palace and Spruce Tree House. The south end of the floor of Balcony House is an ancient sea bottom bedding plane covered with ripple marks.

Unit 4. Loop Road to the Mesa Top Ruins.

Pit houses and pueblos are located on the top of the mesa. The red soil is characteristic of the area and is composed of a fine wind blown dust called loess. It covers not only the tops of the mesas, but most of the flat plains below the Mesa Verde. The rich soil holds moisture well and was not only useful to the Anasazi, but is still important to modern farmers. This dust has been accumulating for probably the last million years and is still being blown in from the southwest today.

Navajo Canyon Overlook.

From this viewpoint we can summarize the geology of this loop road. Directly west, Echo House is in the upper massive sandstone of the Cliff House Formation. The shale break to the left is visible as a thin line about the upper level of the tree line (Figure 16). The slope directly below Echo House is composed of thinner bedded sandstone and alternating shaly layers. Below this slope, the lower massive sandstone has a small cliff dwelling. A shale break makes this alcove possible. Down the

Figure 16. Echo House from Navajo Canyon overlook. Upper Cliff House sandstone (UCH); shale break (SH); lower Cliff House sandstone (LCH); Menefee Formation (M).

cliff, at the base of the yellow sandstone, is the soft gray shale of the upper Menefee Formation. This soft shale is easily eroded and forms very steep slopes below the cliff forming sandstone. South are several layers of lighter colored sandstone below the dark shale. These are sandstones within the Menefee Formation, much softer than the Cliff House sandstone and not continuous. Here, one can compare the typical continuous massive marine sandstones laid down in fairly shallow water near the shore of a great transgressive sea, with the discontinuous sands of the Menefee laid down in great, slow moving, meandering streams along the coastal plain bordering the sea.

No major alcoves sheltering cliff dwellings are located at the base of the Cliff House Formation even though ideal conditions for alcove formation seem to exist with a massive sandstone in contact with impermeable shale. Along that contact are alcoves in various developmental stages in the sandstone, but the underlying shales are too unstable to make solid floors, and they erode rapidly as soon as the protective overlying sandstone is gone.

Unit 5. Wetherill Mesa Road.

The geology of the Wetherill Mesa Road is generally similar to that described in the road log of Chapin Mesa. The road from Far View westward follows along the North Rim of the mesa around the head of the Navajo Canyon tributaries. Occasional views to the north and west show the San Juan Mountains on the horizon about 40 miles to the north, the Sleeping Ute Mountain 12 miles to the west, and on clear days, the Abajo Mountains about 70 miles to the northwest in Utah. The steep cliffs of the North Rim show the solid light colored Point Lookout sandstone capping the softer gray Mancos shales. Overlying the Point Lookout and eroded back from the cliff front, is the soft nonmarine Menefee Formation. The rounded hill tops are capped by the orange-brown Cliff House sandstone. This sandstone is the rock horizon in which Cliff Palace, Spruce Tree House and hundreds of other cliff dwellings are found in alcoves farther south in the Park. Here in the north part of the Park, however, these layers of rock are much softer and do not form cliffs. In this part of the Park the Cliff House Formation was deposited farther offshore in slightly deeper water, and the sediments are distinctly finer as a result.

The road leaves the North Rim and turns south along the long finger of Wetherill Mesa. The view to the south and the southeast shows a very flat mesa top, deeply dissected by canyons. To the south, massive cliff forming yellow cap rock of the Cliff House Formation becomes obvious again. In the far distance on clear days, Shiprock, a volcanic neck about 35 miles to the south, is visible rising abruptly about 1400 feet above the broad flat plain below Mesa Verde.

At this northern end, Wetherill Mesa has been eroded to a very narrow ridge by tributaries of Long Canyon on the east and Rock Canyon on the west. At about the area of the parking lot, the mesa top widens. Many mesa top and cliff dwellings are present on and surrounding the Wetherill Mesa. At the present time two large cliff dwellings, Long House and Step House may be visited on this mesa.

Long House.

Long House was built in an alcove on the west side of Wetherill Mesa at the head of a tributary to Rock Canyon. The alcove is about 100

Figure 17. Ripple marked sandstone overlain by bentonitic shale at rear of upper alcove of Long House.

feet below the mesa top and nearly 500 feet above the canyon bottom. This alcove was formed at the contact of the massive upper Cliff House sandstone and a persistent, thin, impermeable shale layer. Rock Canyon has eroded down through the upper Cliff House Formation, the middle sandy shale and the lower Cliff House sandstone and through much of the soft Menefee shales.

Long House is built on two levels. The outer shallow alcove shelters most of the rooms, and on a level above the main floor, reached by a short ladder, is another, deeper alcove. The two levels are the result of two different shale layers with somewhat different composition. In the back of the upper, inner part of the cave, well preserved ripple marks cover the floor. Immediately overlying the ripple marked sandstone (Figure 17) are thin beds of very soft bentonitic shale. Water seeps out at this level and small water collecting pits, some with feeder grooves (Figure 18), have been pecked out by the Anasazi, indicating that this seep was an important water source at the time of the prehistoric occupation. The shale along which the water seeps contains much bentonite, an altered volcanic ash. This bentonitic shale has the property of swelling greatly when it is in contact with water. The swelling, of course, causes

Figure 18. Water collecting pits in Long House.

Figure 19.   Lower shale layer over floor of outer alcove of Long House.

pressure on the overlying rocks and helps speed up the disintegration of the sandstone at the contact and consequently facilitates the more rapid deepening of the alcove. Many of the deeper alcoves of the Park have such a bentonite layer just above the floor level.

Along the trail near the entrance to Long House, a lower shale layer is visible (Figure 19). Little bentonite is present in this lower shale. Weathering and erosion at this shale level, in conjunction with the sheet spalling from frost, wind and water action have produced the outer alcove.

Step House.

Step House is situated on the east side of Wetherill Mesa at the head of a small tributary to Long Canyon. The cliff dwelling is in a relatively shallow alcove only about 50 feet deep. In contrast with most of the alcoves sheltering major cliff dwellings, this one does not have the typical shale layer and seep at its floor base. The alcove seems to be partially the result of water cascading over the lip of the canyon during heavy runoff periods. Wind erosion plus the continuous sapping and spalling from mechanical and chemical weathering, especially along joints in the massive sandstone, continually enlarges the shelter.

Part II
The Setting

The Mesa Verde is a land of broad, flat topped mesas, cut by deep, sharp shouldered canyons. The northern face of the mesa rises abruptly nearly 2,000 feet above the flat plain near Cortez, Colorado. The top of the mesa slopes gently southward from about 8400 feet elevation at Point Lookout to 6800 feet on Chapin Mesa near the south edge of the Park, a slope of approximately 150 feet per mile. This gentle slope is especially visible from the Far View area where the entrance road reaches the top of the mesa. Here the view to the south looks like a great flat table land. Not so easily seen are the deep, steep walled canyons which cut the mesa into long slender fingers. The drainage of the mesa is mainly to the south, and the major canyons within the Park (Morefield, Prater, Moccasin, School Section, Soda, Navajo and Long) have cut headward until they have reached or nearly reached the North Rim. This drainage pattern is apparent on the geologic map of the Park (Plate A). The long north-south canyons make it very difficult to travel across the mesa in an east-west direction.

Entrance to the Park is from Highway 160 across the low flat plain. The road soon climbs the north face of the mesa, and gains about 800 feet in some three miles of switchbacks along the north face. The campground is located at the head of Morefield Canyon where the Morefield has broached the cap rock of the north face of the mesa and created a broad flat area. Small gullies are eroding the soft face of the mesa from the north also, so that the remaining divide is being attacked from both sides. The road now crosses Morefield Canyon, goes through a tunnel, then comes out into Prater Canyon, another of the north-south drainages of the Mesa. Prater also has cut headward and broached the north face. The road crosses Prater Canyon near the head of the canyon where the resistant sandstone cap has been eroded away and the slopes are gentle. The road follows along the west side of the canyon, climbing gently until it crosses to the north face of the mesa again at the head of Moccasin Canyon. For about the next four miles the road contours along the north face then winds around the head of Little Soda Canyon and to Far View at the top of the mesa at an elevation of a little over 8000 feet. The road now stays on top of one of the mesa fingers, Chapin Mesa, gradually sloping down to an elevation of about 7000 feet at the headquarters and museum. At Far View, another road leads off to the west to Wetherill

Mesa. This road parallels the North Face also, and follows along at the heads of Navajo and Long Canyons. A great deal of mileage is added to avoid the deep canyons. The road finally reaches Wetherill Mesa and continues south on another finger of table land.

The Early Geologic History

The origin of these beautiful mesas and canyons can best be explained by describing at least a little of what has happened here in the last two and a half billion years. Although the oldest rocks which we can see within the Park date back about 91 million years, much older rocks underlie the area. They are known from rocks exposed outside the Park to the northwest and southeast and also from water, oil and gas well records of the surrounding area.

Geologists divide rocks roughly into three major types according to their origin: sedimentary, igneous and metamorphic. Sedimentary rocks, that is rocks formed by the weathering and disintegration of older pre-existing rocks, underlie almost all of the Mesa Verde. Sand, mud, silt and gravel were deposited in stream beds, in swamps, lakes, along beaches and in the shallow sea, and after compaction and/or cementation became sandstone, shale, siltstone and conglomerate. Limestone, composed of calcium carbonate, was precipitated chemically or by the action of organisms in the sea or lakes. Coal was formed of organic material in lush swamps.

There are very few igneous rocks in place in Mesa Verde National Park. They were solidified from liquid molten rock intruded upward along weak areas of joints and cracks in the older rock and are referred to as dikes. Layers of altered volcanic ash called bentonites are also found in many parts of the Park.

The third major type of rock, the metamorphic, is formed from any other pre-existing rock which has been subjected to heat and pressure, usually at great depths below the earth's surface. No metamorphic rocks are found in place within the Park boundaries, but pebbles of both metamorphic and igneous rocks are found on top of the mesas and high on valley terrace levels where they were deposited by streams flowing south from the La Plata and San Juan Mountains a long time ago.

We should digress here to speak very briefly about the geologic age names we must use. Geologists dealing with rocks millions of years old have had to come up with some kind of terminology to describe the age of a rock so that other geologists will know what time frame they are talking about. It never was practical to refer to rocks by their age in years. For one thing, who can comprehend the time span of 50 million years or 84 million, etc. The reasonably accurate dating of geologic time is a very recent and highly technical development. In the early 1800's scientists in Europe began studying rock columns and devised a relative time table which essentially is the time table used today throughout the world. There have been a few additions, refinements and many revisions in the assignments of dates in years, but most of the names remain and are used throughout the world. Geologic time (Figure 2, page 6) is divided into eras and periods. The greatest divisions, eras, are from oldest to youngest: The Precambrian (meaning before Cambrian time) is characterized by very, very old rocks having few indications of the life of the time; Paleozoic (paleos=ancient, zoa=life), Mesozoic (middle life), Cenozoic (modern life). Each of these has been subdivided into periods as shown on the chart, mostly on the basis of major changes in life forms. Further subdivisions have been made, but they are of little interest here.

The rocks on the surface at Mesa Verde are between 78 and 91 million years old, deposited in the great Cretaceous inland sea. The beginning of the geologic history of the area dates back perhaps two and a half billion years. The early background of the pre Cretaceous geology provides a more complete regional understanding (Figure 1, page 5). These older rocks have a place in the history of the Park, even though they are not seen on the surface. We begin with the oldest known, the Precambrian, some two billion years old. These rocks, now deep below the land surface, consist of quartzites (sandstones which have been subjected to great heat and pressure and thus metamorphosed), schists (metamorphosed shales), and igneous granite. No fossils are known in these very old rocks, but the sandstones and shales were probably deposited in a very ancient sea (Figure 20). Subsequently, these rocks were squeezed and bent and broken with attendant intrusions of granite, and mountains were formed (Figure 21).

As soon as the land was raised above sea level, the rocks were subjected to rain and wind, heated by the sun and frozen by frosts. They were broken down into sand grains and pebbles, mud and dust, and the pieces carried by streams, were deposited again across flood plains or in another sea. These recurrent cycles of uplift of the land and immediate subjection to weathering and erosion still continue. The land has been worn back down to a plain or covered by a sea repeatedly. Resultant new deposits of sediments have been laid down in the sea or on flood plains. Another uplift and the repetition of the wearing away process goes on and on. Geologic processes are usually very slow and hardly noticeable in most places, but we have all seen gullies formed or rapidly widened and deepened during a heavy storm. In geologic time, we are

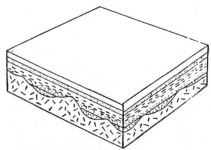

Figure 20. Block diagram of deposition in the Precambrian sea.

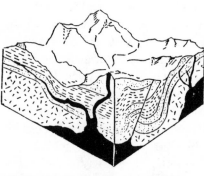

Figure 21. Block diagram of folding, faulting igneous intrusion and erosion of Precambrian rocks.

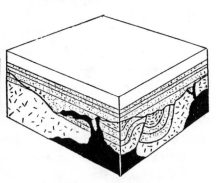

Figure 22. Block diagram of sandstone deposition in Cambrian sea on old Precambrian erosion surface.

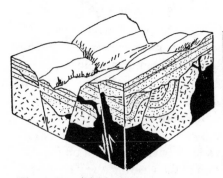

Figure 23. Block diagram of post-Cambrian block faulting and erosion.

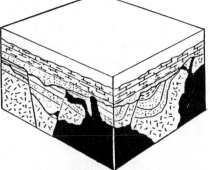

Figure 24. Block diagram of deposition in Devonian and Mississippian sea on erosion surface.

dealing in thousands or millions of years which provides plenty of time for the day by day wearing, softening and breaking down of rocks to erode away whole mountain ranges, not just once, but repeatedly.

After the Precambrian mountains were worn away, a sea once more covered the flat land. It is thought that this sea approached from the west and covered much of the southwestern part of Colorado and northwest New Mexico. In this sea were deposited sandstones (Figure 22), now mostly metamorphosed into quartzites. The sand probably came from the wearing down of the higher land in the general area of the modern San Juan Mountains. There has been much question about the age of these rocks, called the Ignacio Formation, because no diagnostic fossils have been found to help date them. From comparisons with other rocks, they are usually considered late Cambrian. The nearest good exposures of the Ignacio Formation are some thirty miles northeast of the Park near Rico, Colorado.

After the deposition of the sands of the Ignacio Formation, the rocks were broken into huge blocks by pressures within the earth, a process called faulting. Some blocks dropped down, others were squeezed up (Figure 23) and were promptly subjected to erosion and weathering. They began to be worn away, so now the only traces of the Ignacio quartzite are preserved in the downthrown blocks. The land remained high and subject to erosion and weathering for a very long time, and we have no record of Ordovician or Silurian times in this area at all.

Sandstones of the Devonian directly overlie the Cambrian Ignacio Formation. These light gray to red sandstones had their source to the east and were deposited on the ancient irregular erosion surface. An upper member of this Elbert Formation was probably deposited on a great tidal flat. The Elbert is composed mostly of green shales and thin-bedded limestones and has been dated by fossil fish.

Over the Elbert Formation are massive dark brown limestones called the Ouray Formation. These limestones were laid down in a great sea, the third time this area was covered by a sea. Fossils of corals, brachiopods, snails, crinoids and microscopic animals help date the rocks as very late Devonian and early Mississippian. The overlying fossiliferous Leadville limestone shows continuing deposits in this Mississippian Sea (Figure 24).

Strong movements of the earth once more raised the surface above sea level, and erosion and solution removed much of the Devonian and

Mississippian deposits here. So in early Pennsylvanian time as a land area again, soils formed in the low lands and the higher areas were worn down as usual. Gradually a shallow sea entered the area through a trough to the south. Cycles of limestone and salt were deposited to form the Hermosa Group of Pennsylvanian deposits. Another strong uplift drained the sea from the area at the end of the period.

The Permian period is characterized by colorful red beds which give rise to spectacular scenery throughout the Southwest. The major uplift in Pennsylvanian and Permian times developed a high land to the northeast that has been referred to as the Ancestral Rocky Mountains (not the Rocky Mountains that you see now). The rapid erosion of this highland area provided much coarse reddish brown loose material which was deposited in broad stream beds and flood plains to form the Cutler Group.

We thus leave the Paleozoic era, the age of ancient life. Ancient primitive amphibians and reptiles probably wandered across these plains. Many of the sediments laid down at this time have been eroded away, leaving a gap in the geologic history, called an unconformity, between the last of the Paleozoic and the first Mesozoic records found here.

The Mesozoic Era

The Mesozoic Era is often called the Age of Dinosaurs and began the time of rapid development and proliferation of land animals. The highly cross-bedded, fine grained, distinctive orange-red Navajo sands were blown in from the southwest by winds and filled the lower areas of the Cutler flood plains, leaving the old higher areas mainly uncovered in early Mesozoic time. The Dolores Formation crops out along the Dolores River above the Navajo Sandstone only about ten miles north of the Mesa Verde Park entrance. These sediments were deposited in a temperate climate in streams, lakes and on flood plains in a large basin that covered parts of Utah, Arizona, New Mexico and Colorado. First sands (sandstones), gravels (conglomerates), then gradually finer grained sands, silts and limy gravels (limestone conglomerates) accumulated. The latter rocks contain many bone fragments and teeth of reptiles. The basin continued to fill with fine sands and muds through Triassic and early Jurassic time. Lack of useful guide fossils makes exact dating of these sediments difficult.

The Jurassic rocks overlying the Dolores Formation were also laid down under continental conditions, many as fine wind blown deposits in the bottoms of small lakes. The presence of major uranium deposits in the Jurassic rocks of the Colorado Plateau created an intense scientific interest in the sediments and resultant detailed studies. Although no Jurassic rocks are visible on the surface in Mesa Verde National Park, deep water wells in the Park have penetrated sands of this age, and some are well exposed in the Dolores River valley a few miles north.

The Jurassic rocks are mainly sandstone and fine siltstone of the Entrada Formation, probably deposited in great sheets of sand in and between dunes and in small lakes and streams. Much of the Entrada is reddish-orange, fine grained, well sorted (most of the grains about the same size), typical wind-blown deposits. It is known that an ancient sea existed at this time to the north and west in central Utah, and these sands were deposited well back from the shoreline. The sandstones are overlain by the muds and sandy muds of the Summerville Formation which was probably laid down under similar conditions in a large inland lake.

The overlying Morrison Formation, of late Jurassic age is an extremely important deposit. Known throughout much of the Western Interior, it is composed of green, dark red, tan, and gray shale, fine to coarse grained sandstone and some gray to white limestone. These sediments were deposited in rivers, lakes and swamps. The climate during Morrison time was tropical and supported abundant plant growth. As a result of this more favorable environment, a large dinosaur population thrived. The Morrison is famous as the source of the most extensive dinosaur and other reptile collections. Also, bones of small primitive mammals have been found associated with the dinosaurs. This area of swamps and broad streams was ideal for the preservation of the animals. After death, if an animal fell into a swamp and was covered with swamp mud, the chances of preservation were much better than those of an animal dying on an open plain where it would be attacked and dismembered by predators. Streams can tear a skeleton apart also and wear the remains to tiny fragments, but occasionally a body is covered by sand and mud quickly enough to preserve it. In many parts of Colorado, Wyoming and Utah outstanding vertebrate fossils from the Morrison Formation have been discovered and studied. Dinosaur National Monument was created to preserve some of these remains.

A thin layer of gravels forming the Burro Canyon Conglomerate overlies the Jurassic Morrison in this area. These coarse sediments were deposited in fast flowing streams spreading out from highlands to the southwest. There is controversy in placing the exact age of Burro Canyon, but the formation is important here in the Park because quartzite and chert pebbles from it were used by some Anasazi for hammerstones and other tools.

A period of erosion followed the deposition of the Burro Canyon conglomerates, and as a result another gap (unconformity) appears in the geologic record.

The Cretaceous Sea.

The land which is now Mesa Verde National Park had been above sea level through the latest part of the Paleozoic Era, and through the Triassic, Jurassic and early Cretaceous Periods of the Mesozoic Era, a time span of some 150 million years. Then one of the greatest invasions of the sea over continental North America began. A sea invaded from the south from what is now known as the Gulf of Mexico and slowly spread north and northwest over the interior of the continent. At the same approximate time, an arctic sea spread south over Alaska and down across northwest Canada. These very shallow seas moved toward each other slowly over millions of years, with minor fluctuations, until finally they met, and what is now known as continental North America was totally divided by a broad seaway.

A paleogeographic map (map reconstructing the geography of a prehistoric time) has been prepared to give some idea of the general position of land and seaway in the Upper Cretaceous at the time of the greatest extent of the sea (Figure 25). This generalized map does not attempt to reflect the more complex problem of the position of the North American continent relative to other continents at this time. In the past twenty years the studies of continental drift in geologic history has fairly well proven that land masses of the present are not in the same relative positions as they were, for instance, in early Paleozoic or early Mesozoic time. By late Cretaceous, however, continents bore a somewhat closer relationship to their modern contours than previously. So, for this guidebook we will draw rough geographic outlines on the basis of the familiar modern maps.

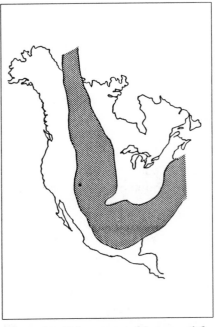

Figure 25. Paleogeographic map of the Upper Cretaceous sea at its greatest extent.

■ Mesa Verde National Park.
▨ Area occupied by the sea.

Dakota Formation.

The first deposit in this inland seaway was about 100 million years ago in early Upper Cretaceous time. As the water encroached over the old, low, relatively flat erosion surface, streams from the west brought in sands and muds. To the west were highlands which during Cretaceous time underwent repeated uplifts, and so furnished sediments to the great downwarping area occupied by the sea.

At first the water was very shallow. The deposits were typical shoreline deposits — beach sands, shallow water cross-bedded shore sands, lagoonal and swamp muds, deltaic sands at the mouths of inflowing streams. The sea did not advance continuously or evenly, and the layers of swamp muds were interbedded with offshore sands as the sea temporarily withdrew and readvanced many times. This sequence of beds makes up the well known Dakota Formation. Near Cortez, coal seams from the old swamps are visible from Highway 160 within the typical brown sandstone sequence. Many of the sandstones show wave ripple marks, tracks and trails of bottom living crustaceans, clams, worms and tubes from burrowing crustaceans. The Dakota sandstone crops out in many places in Colorado and New Mexico, and it or similar formations are known throughout the extent of the great sea area. The sea spread slowly from both the south and the north. Thus, the near shore deposits where the sea first invaded the trough were of necessity deposited earlier in geologic time than the same type of deposits hundreds of miles inland. Therefore, while the same general type of sandstone is found overlying the late Jurassic throughout the Western Interior, it was not all deposited at the same geologic moment. In general, a deposit laid down near shore, closer to the source material from nearby highlands will be coarse, that is, sands and gravels which become sandstones and conglomerates. Deposits far from shore and therefore farther from source materials will be finer, such as muds, eventually forming shales and sandy shales. The Dakota is a near shore deposit of an advancing sea and is predominantly sandstone. It has proven to be a very important aquifer and also a source of gas and oil.

Mancos Formation.

The sea continued to advance, fairly rapidly now. The trough gradually deepened, especially on the west side. The Mesa Verde region was well out from the shoreline which was probably close to the western edge of Utah, some 200 miles away. Now the sediments change from the coarser near shore sandy deposits to fine evenly bedded dark shales. In all, some 2000 feet of Mancos Formation were deposited in quiet offshore conditions. The formation was first described in 1899 from the excellent exposures in the Mancos valley on the east edge of the Park.

Near the Park entrance is the flat eroded surface of the lower part of the Mancos Formation and in the sheer cliffs ahead, the thick soft gray and tan upper Mancos underlying the cap rock of Point Lookout.

Although the Mancos is generally lumped as a great shaly mass, it is not one homogeneous unit. There were shallower and deeper areas in the sea, areas of scouring currents and others of extremely quiet water. The oldest part of the Mancos Formation here, known as the Graneros Member, is mostly barren of fossils, though large limy concretions are common in a zone near the base. Numerous thin bentonites (volcanic ash which has been altered by weathering) appear as thin orange breaks in the dark shale. The bentonite is very plastic and white on a fresh exposure, weathering to a rusty orange color. The presence of bentonites indicates that somewhere at this time there were active volcanoes erupting ash. However, since the ash can be carried in the atmosphere very long distances, just where these volcanoes were located is not known.

An insight into the environment during the Mancos deposition is furnished by the evidence of the life forms in these rocks. The first recognized fossil zone is found some 75 feet above the base of the formation in a unit less than ten feet thick. Within this unit is a bed about a foot and a half thick composed of almost nothing but shell material. The animal whose remains are found in such abundance was an oyster known as *Pycnodonte newberryi* (Figure 26). It was a small oyster with a maximum length of about two inches. The two valves (shells) are unequal. The large (left) valve has a rounded profile with a

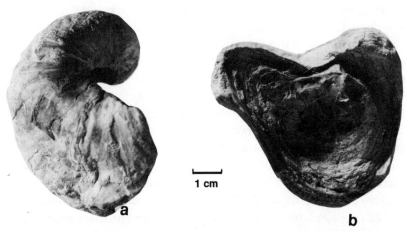

1 cm

Figure 26. Small oysters (*Pycnodonte newberryi*) of lower Mancos Formation: a.) strongly curved left valve; b.) concave right valve in place with left valve.

small flat area near the beak where it was probably attached to the surface of a rock or another shell during a part of its life. The right valve is very small, flat or even concave. This layer of shells makes a very hard, persistent bed which ranchers in the region report as a very difficult layer to get post holes through. The oysters weather out of the layers and litter the slopes below. Also, partially due to the dry climate, the shells do not disintegrate readily. Their almost spherical shape is easily rolled, therefore the fossils accumulate in great numbers in gullies. It is thought that these animals lived in fairly shallow, clean, probably rather warm water. The water was quiet, as evidenced by the fact that most of the shells are intact, not worn by wave or current action. Other fossils found with these oysters include shark teeth and very small snails, indicating a period of slow deposition under quiet conditions, far from shore.

More dark shales overlie these oyster beds. Why did they die out here after thriving in such great colonies? As usual, no one really knows. Perhaps more muds were being brought in by currents, and the animals could not live in the muddy waters. Perhaps the water deepened beyond their life tolerance. Maybe there was a food problem. At any rate, these oysters are an excellent guide fossil because they lived over a fairly large geographical area, but have a very short vertical range geologically. Whenever they are found in place, the geologist knows the age of the particular rock, here dated at about 91 million years old.

The water deepened. Less and less mud reached the area. Finally a limestone, known as the Greenhorn Member of the Mancos Formation, was deposited. Now as the sea reached its greatest advance the waters cleared and again warmed. Studies have shown that there was an abrupt faunal change at this time from warm temperate to partially subtropical temperature.

There is a rather limited fauna of thin flat clams, small oysters and fish scales in the Mesa Verde area. East of here, near Pueblo, Colorado, this Greenhorn interval is represented by very rich animal assemblages and thicker beds of limestone. Perhaps the Mesa Verde area was partially cut off from the main body of the sea by currents or some other barrier. The evidence for whatever caused the difference has been lost because of erosion as the modern Rocky Mountains were elevated some 30 million years later. However, the Greenhorn limestone sequence here consists of only about 15 feet of limestones and thin limy shales. Some of the limestones are fairly solid and contain fine pyrite and pyritized fossil clams. The tops of many small hills from the Mesa Verde north to Dolores and westward to Dove Creek are capped by this blue gray blocky limestone.

The sea reached its greatest extent and then began a slow regression. Conditions became similar to those preceding the period of limestone deposition. Muds became more abundant, and the deposition of the Carlile shales began. The shoreline retreated slowly eastward and

Figure 27. Ammonite (*Collignoniceras woollgari*) impressions in the lower Mancos Formation: a.) juvenile; b.) young adult.

northward. Limestone gave way to limy shales, even bedded, generally without fossils or indications of bottom dwelling organisms. It has been suggested that this barren condition was due to an oxygen-poor sea bottom. Locally, however, at this horizon in a deposit of very thin-bedded brown and gray banded shales, there is a tremendous accumulation of juvenile ammonites (*Collignoniceras woollgari*) (Figure 27). Ammonites are an extinct form of the mollusc group called Cephalopods. No ammonites still live, but other Cephalopods such as the Chambered or Pearly Nautilus, the squid and octopus in modern waters give us some clues about the ancient extinct ammonites which ruled the seas in Cretaceous time. Most of the ammonites found in these shales are only one-half inch to an inch in diameter. They are all juvenile individuals of a form which may be over five inches across in the adult stage. This species has worldwide distribution. However, most assemblages have a variety of sizes of individuals included. Another location of only juve-

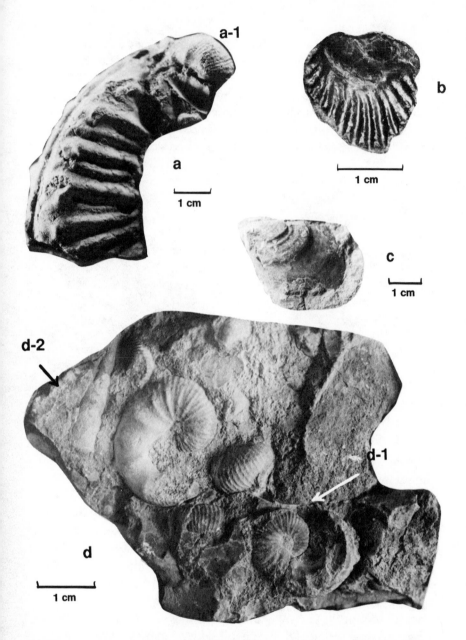

Figure 28. Some typical fossils of the Juana Lopez Member of the Mancos Formation. a.) ammonite fragment (*Prionocyclus wyomingensis*) with a-1) small *Scaphites* ; b.) oyster (*Lopha lugubris*); c.) clam (*Inoceramus dimidius*); d.) slab with assemblage of ammonites; d-1) *Scaphites warreni*, d-2) *Baculites undulatus*. One centimeter scale indicates size of specimen.

nile specimens is found at Black Mesa, Arizona, while near Shiprock, New Mexico, specimens of both adults and juveniles occur. These free swimming animals could range freely from shallow to deep waters. For the most part they were restricted to normal saline waters and required good oxygenation. A striking difference in ornamentation between the juvenile and adult forms may reflect a camouflage adaptation to an environment, and it is entirely possible that the fine ornamentation of the juvenile indicates a different life zone from that of the adults. The accumulation of almost exclusively juvenile ammonites in such quantity has no definite explanation at this point. Other fossils found with the hundreds of young ammonites are small oysters, shark teeth, small clams and barnacle fragments.

Regression of the sea continued with deposition of dark, thin-bedded shales. Discontinuous beds of large brown concretions occur in the shales. Once again thin bentonite layers are seen as orange breaks in the dark shale outcrops, indicating intermittent volcanic activity. Deposition seems to have been more rapid, probably because of an uplift in the marginal source area to the west. Only fragments of fossil clams are found in these shales. The water was probably shallow enough that bottom sediments were reworked by currents and waves, so that the remains of animals living in the sea were broken up.

The shales become sandier with more clams. Finally capping the soft shale and sandy shale sequence are two beds of highly fossiliferous sediments called the Juana Lopez member of the Mancos Formation. These beds look like sandstone, but will completely dissolve in hydrochloric acid. Instead of the typical quartz sand grains, here most of the grains are composed of calcium carbonate. The grains are the very small fragments of sea shells ground to sand grain size by the waves. In this area these beds are rusty brown in color and cap the small hills on the north edge of the Park. This Juana Lopez Member marks the beginning of another transgression of the sea. It must have been a time of high levels of wave and current activity. Large beds of clams and oysters occupied the sea bottom, and free swimming ammonites the water above. In the lower bed the fossils are badly broken, but in the upper bed the individual fossils are better preserved and more complete, perhaps indicating that the zone of wave action had moved a little farther southwest. Figure 28 shows some of the typical fossils of this unit.

The water deepened as this transgression continued and quiet

47

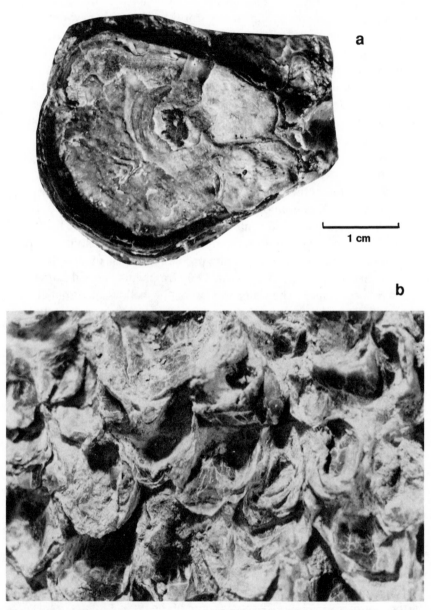

a

1 cm

b

Figure 29. Oysters (*Pseudoperna congesta*) of the Niobrara member of the Mancos Formation. a.) left valve of isolated individual; b.) colony of oysters. One centimeter scale indicates size of specimen.

Figure 30. Large clams in upper middle Mancos Formation. a.) *Inoceramus (Platyceramus) platinus*, large, flat, thin shelled clam; b.) *Inoceramus (Endocostea) simpsoni*, the groove from beak to posterior edge may be the result of a parasite coexisting with the clam, (W. Cobban, personal communication). One centimeter scale indicates size of specimen.

water deposits of thin-bedded dark limy shales followed. Only a few fragmentary clams have been found in these shales.

Once again the sea reached a maximum expanse, and very little sediment reached this part of the basin from the land far to the west. The water was quiet, warm and clear, and widespread oyster beds and very large thick shelled clams covered the bottom of the sea (Figure 29). Some of the clams are as much as eighteen inches across. The oysters are mostly small, from one-half inch to one and a half inches in length, but are present in great numbers encrusting the clams or each other. These beds are correlated with the well known Niobrara Formation in eastern Colorado.

1 cm

Figure 31. Tracks of crustacean (?) in upper Mancos Formation. One centimeter scale indicates size of specimen.

The quiet water oyster beds are overlain by shales and sandy shales as the sea began to retreat again. Large flat clams (Figure 30) are fairly common in these beds. The fossils are not broken up, suggesting that there was little disturbance in the bottom sediments, and deposition was more rapid at this time with more material being brought in from the shore line. However, the land was still quite a distance away.

As the sea retreated further, more and more sand was washed into the trough. Sandstones and sandy shales were deposited. Trails and tracks show that this sea bottom was occupied by many animals: clams, snails, worms and crustaceans. Some of the bedding planes have clear tracks of probable arthropods of various sizes (Figure 31). However, few intact shells have been found because the bottom living organisms destroyed most of the remains before they were buried.

As the water became shallower, more and more bedding planes show ripple marks and trails, tracks and burrows of shallow water animals. The sea continued to retreat, with minor transgressive pulses.

Little remains in the line of fossils. The water was shallow enough that most animal remains were destroyed by wave and current action, but occasional specimens of much broken ammonites and clams are found. This zone of the upper Mancos is mostly yellow-gray sandy shales and shaly sandstones.

This brings to a close the deposition of the Mancos Formation. It took about ten million years for some 2000 feet, primarily of shales, to be deposited. As has been shown, it was not an even, regular deposition. Some intervals received sediments slowly, and at other times it would seem that the source area must have been uplifted and much more sediment carried into the seaway. Although the great gray, much gullied slopes on the north side of the Park look like a fairly homogeneous unit, in reality quite a variety of environments existed during the ten million years of sea history.

Mesaverde Group.

Overlying the thick Mancos Formation is the Mesaverde Group of formations, originally named and described in 1875, long before the area became a National Park. This group is subdivided into three formations, from the oldest, the Point Lookout, to the Menefee, to the Cliff House which is the youngest.

Point Lookout Formation

Directly overlying and grading into the Mancos Formation is the Point Lookout Sandstone. The contact between the two formations is so gradational that it has been traditionally placed at the level where the rocks become more sandy than shaly. This kind of transitional contact is obviously very difficult to pinpoint in the field. A more logical proposed definition places the contact at the base of the first massive sandstone. The transitional zone is well exposed in the road cut at the top of the pass which drops down from the north into Morefield Valley above the campground.

There was continued withdrawal of the sea and the water became more and more shallow, the sediments coarser and coarser, deposition increasingly more rapid. There were temporary shoreline fluctuations so that sandy shales are interbedded with thin sheets of sandstone. But

Figure 32. Alcove in Point Lookout Formation in Prater Canyon.

sandy sediments predominate.

As the sea retreated to the northeast, sands were dumped into this area by large rivers flowing in from the southwest. The water was shallow. Ripple marks, trails of bottom living animals such as clams,

Figure 33. Small alcoves forming in upper Point Lookout Formation.

worms and crustaceans are common on the sandy surfaces. Finally thick massive sheets of sandstones were deposited. These sediments are now the cliff forming beds, each 30 or 40 feet thick, which form the resistant cap rock along the north face of the Mesa Verde. Their resistance to erosion has protected the soft underlying Mancos shales from the rapid erosion which takes place as soon as the cap rock is removed. In most places within the Park there are two major massive sandstones present in the Point Lookout Formation, with shales and shaly sandstone beds between. An eight inch bentonite layer underlies the lower sandstone, indicating renewed volcanic activity. Small alcoves have been weathered near the top of the upper massive sandstone, and some of these have been occupied by the Anasazi (Figure 32). Most of the alcoves are too small (Figure 33) for such use.

On some of the surfaces of the massive sandstones are tracks and trails of indeterminate animals, bits of driftwood and occasional shells of clams and ammonites. Because of wave and current action, few well preserved fossils are found in this formation. Shallow water deposits near the shore where large amounts of sand are being washed into the sea from nearby highlands are not ideal for the preservation of fossils. Waves beat up shells, and bottom feeding scavengers work through the sands and further disintegrate any remains.

Menefee Formation.

The sea finally withdrew from this part of the trough. The new shoreline was only a little way north and east. A broad, almost flat, coastal plain which had been the sea bottom emerged. Sluggish streams meandered over the area draining slowly toward the sea to the north. Broad shallow swamps formed and dark muds and woody material accumulated. The climate was warm and wet, and lush vegetation developed along the streams and on the interstream areas. The swamps accumulated enough plant and woody material to give rise to coal beds eventually. Coal beds in the Park are relatively thin, not more than one to two feet thick and mostly much thinner, but outside the Park they are considerably thicker and are mineable. Between the coal beds are dark shales and brown woody shales, the ancient soils of this great low lying plain.

The contact between the marine Point Lookout Sandstone and the continental Menefee Formation is easily recognized here at the Park. The

Figure 34. Top of Point Lookout sandstone with typical concretions.

top sandstone layer of the Point Lookout weathers rusty brown and has large, iron-rich, dark concretions in it (Figure 34). This sandstone is also frequently covered with ripple marks. The lower layers of the Menefee are very dark brown or black soft shales with thin beds of shiny black coal. In road cuts it is a problem to keep the loose pieces of these soft rocks off the highway. There are thin sandstone beds from an inch to a foot thick interbedded with the shales, probably representing storm deposition when more loose sands were washed down from the higher lands to the southwest. Within some of these sands are excellent leaf impressions. Many of the leaves closely resemble modern broad-leafed plants related to the laurel, witchhazel and camellia families. Impressions of palm fronds and conifers are also found in the sandstones (Figure 35). At this time not enough detailed study of Cretaceous plants has been done by paleobotanists to allow specific identification of these leaves, but by inference with similar flora in the modern world, we can postulate that the climate was warm, subtropical probably, without vast seasonal temperature changes and with plenty of moisture. One might envision broad, lush, flat lands with heavy vegetation, tall grasses, flowering plants and many trees. Very few reptile bones have been recovered from these beds, but with as much vegetation as was present it would be logical to assume that there must have been a population of large and small dinosaurs and other reptiles, as well as small primitive mammals.

Gradually, more and more sandstone beds were deposited with less and less woody material. The middle unit of the Menefee consists of a series of highly cross-bedded, irregular, thick sandstones, interbedded

Figure 35. Some typical fossil plant fragments of the Menefee Formation: a/b.) palm frond fragments; c.) conifer; d/e.) dicotyledon leaf imprints. One centimeter scale indicates size of specimen.

55

Figure 36. Cross-bedding in middle of Menefee sandstone.

with shaly sandstones. These thick sandstones are much softer than the Point Lookout sandstones and do not form spectacular cliffs. However, they are more resistant than the Menefee shales and coals, and crop out on the canyon sides as white rounded rock ledges. These sandstones were deposited in large broad streams and show the irregular stream bottom shape. In contrast with the marine sands described previously, no individual bed can be traced very far. The sandstones are highly cross-bedded in places (Figure 36), irregular bedded overall, with massive and thin beds intertonguing. Within the Park there are two or three of these massive sandstone units, each 20 or 30 feet thick, occupying the same general position within the Menefee. It is a zone of sandstones deposited by great rivers bringing in much sand from the southwest. There must have been renewed uplift in the source area to furnish such large quantities of sand again. Within the zone of sandstone are layers of bentonite (altered volcanic ash), once again indicating volcanic activity. Within this middle Menefee sequence there are not only numerous thin bentonites, but also a very persistent thick layer of solid bentonite as much as three feet thick in places (Figure 37). Some bentonites have the peculiar characteristic of absorbing water readily and swelling to many times their original mass. As a result, when it comes in contact with water, this seemingly solid rock takes on moisture, begins to swell, and becomes very sticky and soft. This thickest layer of bentonite underlies one of the main Menefee sandstones. Where the sandstone is still in place the bentonite is protected from moisture and remains a solid rock as shown in Figure 37, but where the sandstone has been eroded away, the bentonite softens rapidly. Figure 38 shows an area in the bottom of Little Soda Canyon where animals have walked in a bentonite mush and sunk down five or six inches. In most of the major canyons, white or light gray washes of weathered bentonites are visible.

The presence of such large amounts of bentonite within the sandy Menefee suggests some very active volcanoes. At a number of places in the Park large broken tree trunks (Figure 39) are found in the thick of volcanic ash beds. These tree trunks have a diameter up to 33 inches and although preservation is not good enough for exact identification, they are probably some kind of pine. So far, none of the trees has been found in a living position. All are broken and apparently carried into valleys and buried by ash and mud flows. A few fragments of reptile bone have been found in these beds. Many leaf imprints are found in the bentonite

Figure 37. Solid unweathered bentonite A) protected by sandstone B) in middle Menefee Formation.

and in sandstones surrounding it. In the northwest part of the Park barite crystals are found in concretions in this middle sandy zone of the Menefee. Elsewhere in the Park yellow calcite crystals are found in the concretions at this horizon.

Figure 38. Bentonite weathered into a soft sticky, light gray clay.

After this period of volcanic activity things seem to have settled down again. Once more brown shales and coals were deposited as broad coastal plain deposits adjacent to the sea, which was still north and east of the Park. Thin sandstones are interbedded with the coals and shales, but this upper unit of the Menefee closely resembles the lower unit. Near the top are more and more soft sandstones, as this continental phase of deposition drew to a close, and the land once again began to be covered by the sea.

Figure 39. Fossil tree trunks in middle Menefee Formation.

Cliff House Formation

The contact between the Menefee and the Cliff House is another gradational one here at the Park. The sea gradually lapped farther southwest again. The shoreline fluctuated repeatedly so that there are fine sands and reworked muds from the beach, interbedded with the swamp and lagoonal deposits. Eventually the sea won, and the area once again became part of the broad, shallow, marine basin. The contact between the two formations is usually placed just above the level where the last coal appears. Obviously this particular horizon will vary from place to place with an advancing sea over an old low, relief surface.

The Cliff House Formation takes its name from the presence of the famous Anasazi cliff dwellings in the alcoves and niches weathered in the sandstones. The formation typically consists of two massive, cliff forming sandstone beds, each over 100 feet thick, with thin shale and sandy shale partings between them. The upper of these sandstone beds is the member most frequently, although not exclusively, inhabited by the Anasazi around Chapin Mesa. Toward the north part of the Park, the upper sandstones become more shaly and less resistant to erosion and are not cliff formers.

The Cliff House Formation has a fairly typical orange-buff color. Many of the major sandstones are cross bedded deposits formed along barrier beaches of a rapidly transgressing sea. The sand grains are of fairly even size as a result of winnowing and reworking by waves and currents. Not very many good fossils are found within the Cliff House because of its environment of deposition. There were undoubtedly many animals living in and near the sea, but most of their remains were destroyed before fossilization was possible. Shallow water with vigorous wave and current action, plus the ravages of bottom dwelling scavengers leave little to be preserved as good fossils. Many surfaces of the sandstones show clear ripple marks. In some of the top layers of somewhat shalier sandstone a great many burrow fillings are found. The burrows were originally occupied by a crustacean probably resembling a modern mud shrimp. Along with the burrows are many straight ammonites called *Baculites maclearni*. Most of the ammonites are found as short broken sections, but they are recognizable, and are a very useful guide fossil for age determination. In these upper shaly sandstones occasional concentrations of fossils are found. At the time an addition was being built on the Park Museum in 1934, the excavation for the foundation yielded some excellent fossils. Many invertebrates (Figures 40 and 41), clams such as the highly ornamented *Ethmocardium*, Inoceramids, and other clams and oysters, various large and small snails, ammonites such as the straight *Baculites*, the tightly coiled disc-shaped *Placenticeras*, a starfish and sea urchins, have been collected from this horizon in various parts of the Park. The starfish (Figure 41e) is most unusual. In the Western Interior only a few fragments of starfish have been found in Wyoming and Texas in rocks of this age. So this excellent impression of an almost complete specimen is extremely rare and inter-

Figure 40. Some typical fossil clams of the Cliff House Formation: a.) *Inoceramus*;
b.) *Dosinopsis*; c.) *Cymbophora*; d.) *Ethmocardium*; e.) *Modiolus*. One centimeter
scale indicates size of specimen.

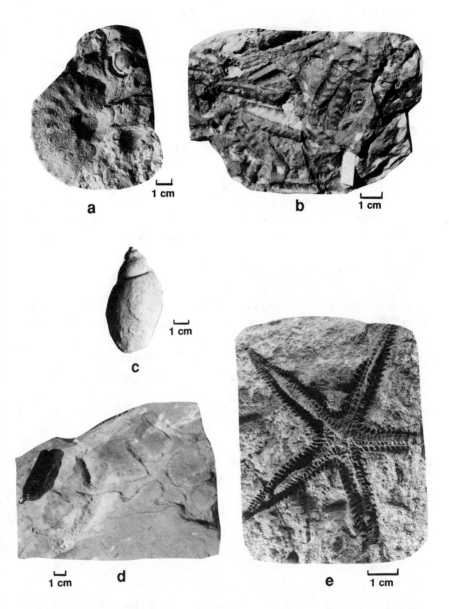

Figure 41. Invertebrate Fossils of Cliff House Formation: a.) tightly coiled, disc-shaped ammonites (*Placenticeras*); b.) slab with numerous straight ammonites (*Baculites maclearni*); c.) snail; d.) Crustacean burrows (*Ophiomorpha*); e.) starfish. One centimeter scale indicates size of specimen.

esting. However, at the time this fossil was found some years ago, no record was made of exactly where or in what rock it occurred. The enclosing rock of the specimen physically resembles the typical Cliff House sandstone, which is probably the source material. However, without a known location, no paleontologist can accept this starfish as a scientifically recognized occurrence. So this beautiful specimen, although interesting, is of questionable scientific value.

Vertebrate fossils are also found in the sandstones. Many shark teeth, large and small fish teeth and bones, amphibians, turtles, Mosasaur, Plesiosaur and other reptile teeth and bones are found in these beds (Figure 42). The water was shallow and this area was close to shore. The mixture of fossilized land and sea creatures proves that this area was very rich in animal life.

The sea continued to advance to the southwest. The massive sandstones can be traced continuously well into New Mexico. Chaco Cultural National Historical Park is about 80 miles southeast of Mesa Verde, and the Cliff House Formation is well exposed there. However, certain fossils found in the Cliff House Formation at Chaco Canyon prove to be about two million years younger than those found at Mesa Verde. The Mesa Verde form has an estimated age of 78 million years, and the Chaco Canyon form is about two million years younger. So, although the sandstone looks the same, and certainly was deposited under the same conditions close to the shore of the same transgressing sea, the sea took about two million years to get from Mesa Verde to Chaco Canyon. By the time the shoreline had reached Chaco, the sea was depositing thick, soft, gray shales offshore over Mesa Verde.

Deposition of the Mesaverde Group was much more rapid than that of the Mancos Formation. A total of about 1200 feet of rock in the Park, including the marine regressive Point Lookout, the continental Menefee and the marine transgressive Cliff House, was deposited in about three million years. This rate of accumulation compared to more than 10 million years for deposition of 2000 feet of Mancos indicates that the highlands which furnished the loose material to the area must have been much nearer and higher than in Mancos time. The quantity of coarse sediments also suggests repeated uplifts of the source area to furnish so much sand. The presence of significant amounts of bentonite indicates volcanic activity, but because volcanic ash can be carried in the atmosphere for very long distances, no conclusions as to the source of this material can be made at this time.

Figure 42. Some vertebrate fossils of Cliff House Formation: a.) tooth of large fish (*Enchodus*); b.) gill cover of large fish (*Enchodus*); c.) shark tooth (*Scapanorhyncus*); d.) rib of Plesiosaur or dinosaur (note impressions of small clams also). One centimeter scale indicates size of specimen.

The Cliff House sandstone is the top layer of rock seen at Mesa Verde National Park today, but it is not the end of the geologic story. The sea continued to advance. As was mentioned, it got down to Chaco Canyon about two million years after it covered Mesa Verde (that is a rate of advance of about two inches per year). Now instead of shoreline sands, the Mesa Verde area was receiving offshore muds and sandy muds. These sediments were compacted to form the Lewis Shale Formation. Once more the sea began to retreat, and another beach deposit similar to the Point Lookout was formed. It is known as the Pictured Cliffs Sandstone in New Mexico. Still more sandstones and shales of the Fruitland and Kirtland Formations were deposited behind the retreating sea. However, none of these deposits remain at Mesa Verde today. Everything above the Cliff House Formation has been eroded away, and that includes a thickness of at least 1500 feet of late Cretaceous sediments. There is little doubt that at one time these Cliff House sandstones were buried more than 1500 feet below the surface.

The Laramide Revolution.

So, after some 20 million years of occupation by the great Cretaceous seaway, the water receded for the last time (no predictions are intended concerning the future). Large scale earth movements began which completely changed the appearance of the region.

In the very latest part of Cretaceous time, widespread uplifts and accompanying volcanism began and continued into the early Tertiary period. Known as the Laramide Revolution, it marked the beginning of the development of the modern Rocky Mountains. The La Plata Mountains and the Sleeping Ute Mountains were formed at this time also. Radiometric dating suggests that these earth movements may have begun about 66 to 70 million years ago in this general area and continued intermittently at least 25 million years. Probably early uplifts consisted of simple bowing up of the area. Later movements have largely obscured the early history, but more or less continuous deformation of the rocks occurred, along with very deep magma (liquid rock) intrusions and surface volcanoes. The rocks were bent and broken and cut by igneous intrusions. The land was rising in some areas and sinking in others. Erosion continued to wear down the highlands and deposit the debris in the low basins. It is thought that the climate at this time was warm and humid, perhaps subtropical, which speeds up the weathering and erosion processes.

Once more the entire Southern Rocky Mountain area was eroded back down to a flat plain in early Tertiary time. There were some local highlands left, but the erosion pretty much kept pace with the uplifts. This level plain can be traced across hundreds of miles of the southern Rocky Mountains and adjacent Great Plains.

Mid Tertiary Volcanism

This period of erosion was followed by another period of volcanic activity extending from about 35 to 26 million years ago in early mid Tertiary time. The present San Juan Mountains northeast of Mesa Verde National Park are the eroded remnants of some of this great volcanic field. These newer volcanics are found covering the earlier Tertiary erosion surface in the San Juan Mountains, thus establishing the age relationship unquestionably.

Volcanic activity continued intermittently in the area. Dikes at Shiprock, New Mexico, and other small igneous bodies were emplaced between 32 and 27 million years ago, mid Tertiary time. In Mesa Verde National Park there are several small igneous dikes (Figure 43). These are the same type of rock as form part of the Shiprock dike, and although they have not been dated here, they are probably of the same approximate age. Usually, as at Shiprock, dikes are more resistant to erosion than the sedimentary rocks into which they have been intruded, and so remain as obvious ridges after the surrounding sediments are worn away. In Mesa Verde National Park differential weathering leaves a somewhat different surface expression. Where the dike exposed in the

Figure 43. Dike in Navajo Canyon. Dike outlined in broken white lines: Cliff House sandstone (CH); Menefee Formation (ME).

Figure 44. Trace of dike (arrow) in Navajo Canyon from the Overflow Parking Area near Park Headquarters at the south end of Chapin Mesa.

lower part of a canyon cut through the soft Menefee Formation, a very low mound or ridge may be visible (Figure 43). However, the trace of the dike through the more resistant Cliff House Formation is usually seen as a sharp sided slot cut into the wall of a canyon (Figure 44), and as a gully across the surface of the mesa top.

The Present Surface

The canyons we see now at Mesa Verde National Park are the result of erosion. The entire area had been worn down to a flat plain with a few more resistant higher areas in early Tertiary time. This erosion plain sloped gently eastward to the La Plata Mountains and northeastward to the San Juan Mountains. Gravel deposits now found on the tops of the mesas include pebbles of rocks which originated in the eroded central parts of both the San Juans and the La Platas. These gravels could only have been left by streams spreading out across the old great plain from those mountain areas long before erosion cut the surface to its present level.

Early in Quaternary time (two or three million years ago), a broad uplift of the entire region once again initiated a new period of active erosion. The mountain regions were probably elevated several thousand feet higher than the surrounding areas. The streams that had been lazily meandering over a flat plain began active downcutting again (rejuvenation of streams). This erosion continues today.

The climate had gradually cooled in these last 20 million years, and glaciers developed in the mountain regions. Although there were no glaciers in the Mesa Verde area proper, streams flowing out from the melting glaciers in the mountains nearby brought some glacial boulders this far. The area was uplifted again, and the streams renewed their downcutting more vigorously, this time cutting new sharp valleys through the slightly wider, older stream beds. Two more periods of glaciation followed with consequent deposits of gravels in stream beds out to the west of the mountains. Since then, the region has been subjected to repeated minor uplifts and rejuvenation of the streams. Most of the Mesa Verde canyons show small terraces which represent periods of stability when the streams cut a little broader valley, and then with rejuvenation the streams cut downward through the old valley floor again. Although there are no permanent, year-round streams flowing in the Park, the intermittent streams occupying the canyons are still in an active downcutting phase. In this climate, extremely rapid erosion occurs during and after each heavy rainstorm, as well as with seasonal snow-melt run off.

Other Surface Features

The alcoves in which the Anasazi dwellings are so well preserved are one of the most important geologic features. After all, beautiful as this country is, it probably would not be a National Park if there were not the great concentration of spectacular prehistoric ruins. The alcoves are the result of a combination of effects.

Many of the alcoves occur at the horizon within the uppermost sandstone of the Cliff House Formation where the sandstone comes in contact with a thin, persistent shale. Ground water percolates through sandstone quite readily because the sandstone is composed of sand grains loosely cemented, and there are tiny openings or pores between the grains through which the water moves. On the other hand, shale is almost impermeable, that is, water cannot move through it. So when it rains or snow melts on the mesa, some of the water soaks into the sandstone and moves downward in the rock until it comes to the shale. Now, since it cannot move down any further, the water seeps along at the top of the shale surface, down the dip or slope of the rock. Eventually, in an area as dissected by canyons as the Mesa Verde, the water seeps out along a canyon wall. These seeps occur at the top of the shale. The moving water has the ability to wear away the sandstone both mechanically and chemically by solution. The sand grains are loosely cemented together by calcium carbonate (lime). Calcium carbonate is very soluble in various acids. The rain water combines with carbon dioxide in the atmosphere and forms a very weak carbonic acid (acid rain did not begin in the twentieth century)! This weak acid slowly dissolves the lime cement. When the cement is dissolved, loose sand grains are left, and the sandstone simply crumbles and is washed away in the next rain storm or blown away by the wind. So the alcoves start as tiny niches in the rocks at the contact of the sandstone with the underlying shale. Slowly the seeping water undermines the sandstone. As the undermining progresses, blocks of sandstone drop down, some just to the floor of the alcove, others tumble down into the canyons. Very gradually between the continuing action of the seeping water, freezing and thawing, and wind and rain storms, some of these alcoves became very large and eventually furnished shelter to the Anasazi. The erosion processes are still going on today, which is one reason continuous stabilization work in the Park is necessary to protect the ruins and alcoves from further natural erosional damage. Of course, the weather-

Figure 45. Thin sheets of sandstone split off of massive cliffs along trail to Step House on Wetherill Mesa.

ing away of the rocks would be much more rapid in a humid climate with more ground water available.

Another factor is involved in the formation of the larger, deeper alcoves. A layer of bentonite up to a foot thick occurs directly above the shale in many of these alcoves. As has been described, bentonite has the characteristic of absorbing water, resulting in a major increase in its volume. This swelling causes pressure at the contact with the overlying sandstone and speeds up the disintegration of the sandstone. It is probably an important factor in the development of the deeper alcoves.

The inhabitants of the alcoves also had seepage water available in the back of some of these shelters. The quantity was probably not great, but any local supply of water would certainly be welcome, and in some of the alcoves small pits have been dug (Figure 18, page 30) to collect the seep water.

Although most of the larger alcoves are the result of weathering and erosion at the contact of the porous sandstone with an impermeable shale, a number of alcoves have no shale or seep present. Usually located at the heads of small side canyons, these alcoves are usually relatively shallow. A combination of weathering and erosional factors are at work in these areas. All the sandstone cliffs along the canyon walls are constantly being attacked by weathering processes, both chemical and physical. As an example, water may evaporate just below the rock surface, and very fine crystals of minerals precipitate. These crystals cause pressure and wedging of very thin slivers of rock away from the main surface of the sandstone. The regular heating and cooling of the surface causes the minerals making up the rocks to expand with the heat and contract when cooled. Repeated over and over, the mineral grains near the rock surface are loosened. The result of these and other surface weathering agents serves to break up the rock so that it can be carried away by winds blowing up the canyons or by runoff water from a gully above the canyon cascading over the cliff edge. Further, you will notice vertical cracks through the massive sandstone layers. These are called joints. Freezing and thawing of water in these joints gradually pries loose sheets of sandstone from the rock face (Figure 45). All of these processes are involved in the formation of the alcoves in the Park.

Ground water has left other marks on the surface of the Park. Rain water falling through the atmosphere forms a very weak acid, as was previously described. As it runs off, the water accumulates in slight natural depressions in the sandstone and gradually sinks into the rock. Once again solution of the lime cement occurs. The shallow depressions deepen first in the center, and then little grooves leading from the center begin to develop as the rapid runoff water filling the depressions dissolves away the softer areas of cement. Eventually, an almost flower-like pattern can develop. Several of these solution rills at different stages of

Figure 46. Development of solution patterns in Cliff House Sandstone on Chapin Mesa.

71

Figure 47. Solution pattern in niche at Sun Temple.

development on the top of a cliff are shown in Figure 46. One of the these solution forms is found enclosed in a niche at Sun Temple (Figure 47). There has been speculation by archaeologists about the significance of its relationship to the Anasazi culture.

Concretions are found throughout the Park in almost all the rocks, but they were not all developed in the same way. Some are much like mud balls with a foreign object like a pebble or fossil shell in the center. These concretions were formed either at the same time the sediments were deposited or while the sediments were being compacted into a rock.

Other concretions are the result of later chemical cementation. Water percolating through sandstone dissolves away minerals. Those minerals can be lime or sometimes iron, which gives the yellow or brown color to the sandstones. If the water moving through the rock comes to a variation in the sandstone, an obstruction like a shell or a pebble or simply a little more solid part of the rock, the water slows down and as a result, drops its load of dissolved material. This dissolved material, lime and/or iron oxide, is precipitated and cements the sand grains into a mass, a concretion. Sometimes there are parts of the sandstones which have less cement than most. When the ground water reaches such an

Figure 48. Concretion in Cliff House sandstone near Park Headquarters.

Figure 49. Detail of round concretion shown in Figure 48.

area, the water pools temporarily, and once again any minerals in the water will be precipitated to fill this area. Many of these concretions are seen in the sandstones of the Cliff House Formation as one walks to the cliff dwellings (Figure 48 and 49). The shapes are variable, depending on the original cause of their formation. Because they are the result of deposition of extra cementing material, they are more resistant to erosion and frequently result in knobby features. Some of these have rather interesting shapes and have been found among the possessions of the Anasazi.

Another type of concretion is common in the middle of the Menefee Formation. These are large, hard, dark brown, iron oxide masses. The centers of some of them are filled with yellowish calcite crystals. Along the north part of the Park some are filled with white barite crystals. Both calcite and barite crystals have been found with other artifacts in Anasazi sites.

Most of the cliffs show streaks of dark brown or black called desert varnish (Figure 50). Gravel and small rocks on exposed parts of the mesa also frequently have a similar thin dark coating. Desert varnish is a thin deposit of manganese and iron oxide and usually occurs on barren cliff faces where the rock surfaces are intermittently wet from runoff water. The origin of this coating has never been totally agreed upon by geologists. One common theory was that the coating was produced by the heat of the sun drawing water held within the sandstone to the surface. When the water then evaporates, the substances dissolved in the water are deposited as oxides on the surface of the rock. However, not enough manganese is present in the sandstones here to furnish the manganese found in these coatings. It is now thought that the manganese and iron are found in dust blown in from the southwest and deposited on the rock faces. In the presence of moisture, bacterial action fixes the manganese and iron particles to the rock surface.

Pieces of onyx marble or travertine are also found among the Anasazi possessions. This brown to tan striped rock is a lime deposit often occurring where water seeps out into a crack in the rock and gradually builds up thin layers of calcium carbonate which may eventually fill the crack.

The bare areas on the tops of many of the cliffs are often weathered into a "turtle back" pattern (Figure 51), a phenomenon which does not have an easy explanation. It is quite likely the result of a combination of both mechanical and chemical factors. It usually occurs on sandstones

Figure 50. Desert varnish streaks on rock wall above Oak Tree House in Fewkes Canyon.

Figure 51. "Turtle Back" weathering pattern above Spruce Tree House.

that are massive and homogeneous and are on exposed mesa tops or cliff edges where there is no protection from sun or rain. It may be that differential heating and cooling, wetting and rapid drying set up stresses which cause very small cracks to develop in the surface of the rock, usually in a more or less polygonal pattern. Once the fine cracks are started, rain water channels along the cracks and causes further deepening. Mechanical and chemical weathering of the surface continues to loosen the sand grains, and rain and wind remove the loose material and deepen the grooves.

Much of the top of the mesa is covered by a fine red soil. This soil is red loess, a fine, wind blown material brought in from the southwest. It has accumulated probably over the last million years, and some is still being deposited. It covers not only these mesas, but most of the flat plain below the Park as well. Wherever it is found, it forms a very rich soil which holds moisture well. Both the Anasazi and modern farmers have found this thick red soil useful for dry farming. Where it washed away from the top of the mesa, agriculture was nearly impossible. After large parts of the mesa were denuded of trees during the Anasazi occupation, erosion removed large amounts of the fertile soil, and may have been one of the important factors in the abandonment of the mesa.

Extinctions

We cannot close a review of the geology of the Mesa Verde National Park without some reference to the great faunal extinctions at the close of the Mesozoic era. During the Mesozoic the seas were dominated by the great variety of ammonites and the land by the dinosaurs. By early Tertiary time both of these major groups of animals had completely disappeared. Why? No one really knows. There are as many theories as there are theorizers, and none completely answers the question.

The ammonites are a large extinct group of quite highly specialized Cephalopods. Modern Cephalopods include the Nautilus with an external shell, the squid with a vestigial internal shell, and the octopus with no shell at all. The ammonites varied from almost microscopic to several feet in diameter and in form from tightly coiled to loosely coiled to straight (Figure 52). Some were streamlined and disc-shaped, others almost spherical, still others coiled like a corkscrew. Many were smooth shelled and others highly ornamented with ribs and spines. Some of these animals crawled on the sea bottom, others swam freely but probably not rapidly, and still others floated at or near the surface. Some were scavengers, others herbivores. Some may have been carnivores. In turn they were prey to the rapidly evolving fish and the marine dinosaurs. Some had very restricted environments (limited by salinity, temperature, oxygenation, depth) in which they could live. Others seemed to occur almost everywhere in the seas. So the old explanation that they became too specialized and so died out with a changing environment is not really adequate. It is known that they began to decline in both variety and number of individuals in early Upper Cretaceous time. So the total extinction some 25 million years later was not the sudden disappearance of a fauna at its peak, but rather the total disappearance of a slowly declining fauna. Perhaps any of the explanations including the role of the major sea regressions, changes in salinity, climate, oxygen levels, food chain problems, new more highly developed predators, increase in cosmic rays, and many other factors, may have played some role. But it seems probable that there was no one overwhelming catastrophic event that marked the end of the ammonite population.

What happened to the dinosaur? Once again many theories and guesses have been published, some serious, some tongue-in-cheek. The theories on the cause of the extinction again are based on climate changes and food changes, oxygen levels, new mammalian predators,

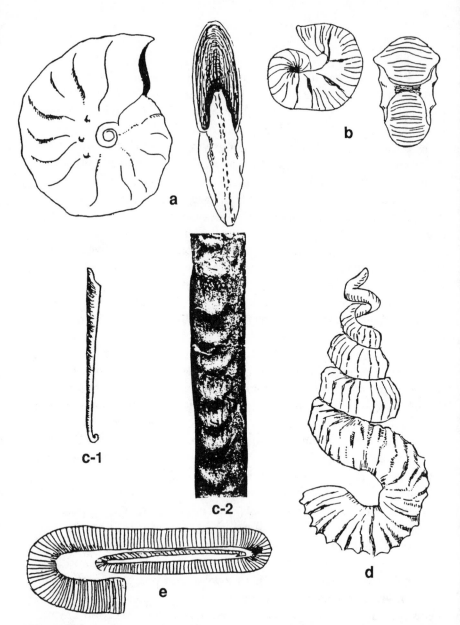

Figure 52. Some ammonite coiling forms in late Upper Cretaceous: a.) tightly coiled, disc-shaped, streamlined ammonite (*Placenticeras* type); b.) loosely coiled, with outer whorl partially uncoiled (*Scaphites* type); c.) straight form (*Baculites* type); c1.) juvenile form showing initial coiling stage and straight shell; c2) typical fragment of adult shell as usually found; d.) corkscrew, helicoid coiling (*Didymoceras* type); e.) irregular coiling (*Ptychoceras* type).

cosmic radiation, changes in gravity and many other factors. Every one of the theories can be refuted for at least parts of the dinosaur population. About 30 years ago someone came up with the idea that at least some dinosaurs were warm blooded, rather than all cold blooded like their reptile relatives, so that environmental temperature might not have been such a controlling factor. Recently this theory has excited the interest of laymen and some scientists again. Evidence of a worldwide dust deposit at the end of the Mesozoic has recently led to speculation of contact with an asteroid at this time. Such an incident could have caused a significant environmental change. Just as with ammonites, dinosaurs came in all sizes from tiny bird size to the huge forms 30 to 40 feet long and weighing as much as 80 tons. Some walked on four legs. Others took to running on two. Many were herbivores and many others carnivores. Certainly these animals also occupied very different environments. Extinction was gradual again, and no one factor seems to account for the complete extinction of an entire fauna. Don't blame man, he did not appear for another 60 million years.

Glossary

Ammonite:

One of an extinct group of molluscs, distantly related to the modern chambered nautilus, squid and octopus. The name refers to the shell shape of some forms which was likened to the horn of the Egyptian God Ammon. Ammonites were the dominant invertebrates in the seas of the Mesozoic Era and had varied shell shapes from straight to tightly coiled in different ways (Figure 52, page 78).

Aquifer:

A water-bearing layer of permeable rock, such as sandstone.

Arthropod:

A large group of animals having jointed appendages, segmented bodies and a well developed head; includes such animals as centipedes, insects, crabs, lobsters, shrimp, spiders, scorpions, ticks, etc.

Barite:

A white, yellow or colorless, very heavy mineral composed of barium sulfate, found in crystals within concretions in the Park.

Bedding plane:

In sedimentary or stratified rocks, the division planes which separate the individual layers or beds.

Bentonite:

A soft, plastic, light colored rock mainly consisting of clays altered from volcanic ash.

Butte:

A small remnant resulting from further erosion of a mesa. A steep sided, flat-topped hill.

Calcite:

A common rock-forming mineral composed of calcium carbonate, as white, yellow or light colored crystals in concretions or as cementing material in sedimentary rocks, easily soluble in hydrochloric acid.

Carnivore: An organism that feeds on animals, living or dead.

Chert: A hard, extremely dense, dull or semiglossy sedimentary rock consisting mainly of silica with crystals too fine to be seen with the naked eye.

Concretion: A hard, compact, rounded or irregular mass of mineral matter.

Conglomerate: A sedimentary rock made up of pebbles or gravel cemented together.

Continental deposit: A sedimentary deposit laid down on land or in bodies of water not directly connected with the ocean (as opposed to marine deposits).

Continental drift: The concept that the continental masses have moved in relation to each other as a result of plate tectonics.

Correlate: To link together strata of the same age in separate locations.

Crop out: To be exposed at the surface.

Cross bedding: An internal arrangement of layers in a stratified rock, characterized by minor beds at an angle to the original surface or principal bedding planes.

Crustacean: Any arthropod characterized chiefly by the presence of two pair of antennae on the head; most are aquatic and breathe by gills; includes such animals as shrimp, water fleas, barnacles, crabs, etc.

Delta: The more or less triangular area of sediment deposited at the mouth of a river where it flows into a sea or lake.

Desert Varnish: A thin dark, hard, shiny or glazed film, coating, stain or polish composed of iron oxide with traces of manganese oxide and silica, formed in deserts after long exposure. Seen as dark stains on the cliffs and as dark shiny coating on gravels in exposed areas on the mesa.

Dike: A tabular igneous intrusion which cuts through the surrounding rock structures.

Dip: The angle between the slope of a bedding plane of sediments and the horizontal.

Ecological niche: The position of an organism or a population in the environment as determined by its needs, contributions, potential and interaction with other organisms.

Ecology: The study of the relationship between organisms and their environment, including the study of communities, patterns of life, natural cycles, relationships of organisms to each other, biogeography and population changes.

Era: A geologic time unit including two or more periods.

Erosion: The general process by which the rock materials of the earth's crust are loosened, dissolved or worn away and moved from one place to another by natural forces.

Fault:	A zone of rock fracture along which there has been displacement.
Fauna:	The entire animal population of a given area, environment, formation or time span.
Flora:	The entire plant population of a given area, environment, formation or time span.
Formation:	A recognizable, named, mappable unit of rock.
Fossil:	The remains, impressions, track or indication of the presence of a prehistoric plant or animal.
Group:	Two or more formations composing a unit.
Herbivore:	An organism that feeds on plants.
Igneous:	A rock solidified from molten material.
Intrusion:	The process of moving magma (liquid rock) into preexisting rocks.
Invertebrate:	Animals which lack a back bone.
Joint:	A break in a rock without displacement.
Laccolith:	An igneous intrusive rock form which has a domed top and flat base and usually causes an up arching of the roof rock.
Laramide Revolution:	A time of deformation and igneous intrusions typically developed in the eastern Rocky Mountains of the United States, whose several phases extended from late Cretaceous until very early Tertiary.
Limestone:	A sedimentary rock consisting chiefly of cal-

cium carbonate deposited chemically and/or organically.

Lithology: Character of a rock.

Loess: A very fine silt or clay originating in arid regions and transported by wind.

Magma: Liquid rock material deep within the earth from which igneous rocks are formed.

Meandering stream: A winding, looping stream pattern produced by a slow moving stream swinging from side to side across a wide flood plain.

Member: A specially developed part of a varied formation, if it has considerable extent.

Mesa: An isolated nearly level land mass standing distinctly above the surrounding country, bounded by abrupt or steeply sloping erosion scarps on all sides and capped by layers of resistant almost horizontal rocks.

Metamorphic rock: Any rock derived from a preexisting rock subjected to natural processes of heat and/or pressure which cause chemical or structural alteration sufficient to produce a new rock type.

Onyx marble: A banded, hard compact generally translucent variety of calcite, usually deposited from cold water solutions, often in caves, or along joints in rocks.

Outcrop: The exposure of bed rock or strata projecting through the overlying cover of detritus or soil.

Paleobotany: The study of plant life of the geologic past.

Paleogeographic map:	A map that shows the reconstructed physical geography at a particular time in the geologic past, including such information as the distribution of land and sea, etc.
Paleontology:	The study of life in past geologic time based on fossils.
Period:	A geologic time unit, a subdivision of an era, the fundamental unit of the standard geologic time scale.
Pyrite:	"Fool's gold", iron sulfide, found in brassy yellow crystals.
Quartzite:	A metamorphic rock consisting mainly of quartz and formed by recrystallization of sandstone.
Regressive sediments:	Sediments deposited during the retreat or withdrawal of a sea from a land area or during the emergence of the land.
Rejuvenation of stream:	The action of stimulating a stream to renewed erosive activity, as by uplift of land mass or a drop of sea level.
Ripple mark:	A very small ridge of sand resembling or suggesting a ripple of water and formed on the bedding surface of a sediment.
Rock column: (stratigraphic)	The vertical or chronologic sequence of geologic formations in a region.
Saline water:	Salt water, such as sea water.
Sandstone:	Sedimentary rock made up of compacted and/ or cemented sand grains.

Schist:
A strongly layered metamorphic rock which can be readily split into thin flakes or slabs because of minerals such as mica being lined up parallel with each other. Schists may be the result of metamorphism of sedimentary shales or of some igneous rocks.

Sedimentary:
Rocks which result from the breaking down of preexisting rocks or by the chemical precipitation from solution or secretion by organisms.

Shale:
A sedimentary rock composed of consolidated muds which split easily parallel to the bedding.

Strata:
Layers of sedimentary rock.

Stratigraphy:
The study of the layers of the rocks of the earth's crust, their structure and relationship to each other, the conditions of their formation and their fossil content.

Stream terrace:
One of a series of level surfaces in a stream valley, more or less parallel to the stream channel, produced during a former stage of erosion or deposition.

Transgressive sea:
The advance or extension of the sea over land areas.

Vertebrate:
Animals which possess a back bone.

Weathering:
The process of disintegration or decomposition of minerals and rocks as a result of exposure to rain, frost, and sun, but with little or no movement of the loosened material from one place to another.

Bibliography

Cobban, W.A. and Scott, G.R., 1972, Stratigraphy and ammonite fauna of
  the Graneros Shale and Greenhorn Limestone near
  Pueblo, Colorado: U.S. Geological Survey Professional
  Paper 645, 108p.

Cobban, W.A. and Hook, S.C., 1979, *Collignoniceras woollgari woollgari*
  (Mantell) ammonite fauna from Upper Cretaceous of
  Western Interior, United States: New Mexico Bureau of
  Mines and Mineral Resources Memoir 37, 51p.

Ekren, E.B. and Houser, F.N., 1965, Geology and petrology of the Ute
  Mountains area, Colorado: U.S. Geological Survey
  Professional Paper 481, 74 p.

Haynes, D.D., Vogel, J.D., and Wyant, D.G., 1972, Geology, structure,
  and Uranium deposits of the Cortez Quadrangle,
  Colorado and Utah: U.S. Geological Survey
  Miscellaneous Investigations Series, Map I-629, scale
  1:250,000, 2 sheets.

Hook, Stephen and Cobban, William A., 1976-77, *Pycnodonte newberryi*
  (Stanton) - Common guide fossil in Upper Cretaceous of
  New Mexico: New Mexico Bureau of Mines and Mineral
  Resources, Annual Report 1976-77, p.48-54.

Kauffman, Erle G., 1977, Geological and Biological Overview: Western
  Interior Cretaceous Basin: The Mountain Geologist, Vol.
  14, Numbers 3 and 4, p. 75-99.

O'Sullivan, R.B., Repenning, C.A., Beaumont, E.C., and Page, H.G., 1972,
  Stratigraphy of the Cretaceous Rocks and the Tertiary
  Ojo Alamo Sandstone, Navajo and Hopi Indian
  Reservations, Arizona, New Mexico, and Utah: U.S.
  Geological Survey Professional Paper 521-E, 65p.

Siemers, Charles T. and King, Norman R., 1974, Macroinvertebrate
  paleoecology of a transgressive marine sandstone, Cliff

House Sandstone (Upper Cretaceous) Chaco Canyon, Northwestern New Mexico: New Mexico Geological Society Guidebook 25th Field Conference, p. 267-277.

Tweto, Ogden, 1980, Tectonic history of Colorado: in Colorado Geology, Rocky Mountain Association of Geologists, p. 2-7.

Wanek, Alexander A., 1959, Geology and fuel resources of the Mesa Verde area Montezuma and La Plata Counties, Colorado: U.S. Geological Survey Bulletin 1072-M, p. 667-717.